Molecular Biology Biochemistry and Biophysics 25

Advanced Methods in Protein Sequence Determination

Edited by

Saul B. Needleman

With Contributions by

R. F. Doolittle · H. Falter · M. J. Horn
K. K. Kannan · R. A. Laursen · G. A. Mross
S. B. Needleman · E. Nieboer · M. Reichlin

With 97 Figures

Springer-Verlag Berlin Heidelberg New York 1977

Professor SAUL B. NEEDLEMAN
Scientific Affairs
Abbott Laboratories
North Chicago, IL. 60064/USA

ISBN-13:978-3-642-81165-4 e-ISBN-13:978-3-642-81163-0
DOI: 10.1007/978-3-642-81163-0

Library of Congress Cataloging in Publication Data. Main entry under title: Advanced methods in protein sequence determination. (Molecular biology, biochemistry, and biophysics; 25) Bibliography: p. Includes index. 1. Proteins — Analysis. I. Needleman, Saul Ben, 1927—. II. Series. [DNLM: 1. Amino acid sequence. 2. Proteins — Analysis. WI MO195T no. 25/QU55 A244]. QD431.A28. 547'.75. 77-22808.

2131/3130-543210

For
Jack and Celia
Abe and Jean

Preface

Confusion now hath made his masterpiece
Macbeth II iii 72
Whence and what are those execrable shape?
Paradise Lost Ib 1 681
Confusion worse confounded
Paradise Lost Ib 1 995

When the manuscript for the first part of this book was proposed, it was anticipated that the discussion of the entire field of protein sequencing could be covered in a single volume — from purification and characterization of the protein through fragmentation by chemical or enzymic means and, finally, to reassembly of the identified individual peptides into the reconstructed total sequence. It soon became evident that this would not be possible. While the intent was to restrict the exposê of procedures only to that information concerned with "hands on" wet chemistry, it became apparent that a thorough presentation would require, in addition, a discussion of certain instrumental and more theoretical approaches not included in the first volume. Furthermore, the entire understanding of the field of protein sequencing has advanced appreciably since the inception of this book.

The purpose of the first volume was to provide practical information in sufficient detail to permit the researcher to undertake the actual sequencing procedures in his own laboratory. The intent of the present volume, however, because of the nature of the techniques in consideration, is, rather, to provide an understanding of the principles and of the language of perhaps more esoteric procedures, not so that the researcher might attempt to undertake these, but rather that he might be able to avoid confusion and be able to discuss intelligently with appropriate specialists the application of these special techniques to the structural problems at hand.

In the first volume, one chapter was devoted to the solid-phase synthesis of small peptides as final proof of structure in areas difficult to resolve by chemical means. The first two chapters of the present volume deal with the application of solid-phase supports to the stepwise degradation of proteins. One chapter deals with the overall procedures involved while the second elaborates on the specific problems encountered in coupling the protein to the solid support and on strategies for resolving these problems.

More and more reports appear regularly in the literature describing N-terminal blocked proteins. The third chapter constitutes an expansion of techniques only alluded to in the first volume for sequencing such proteins which lack the free α-amino group.

Sequencing by immunologic techniques is not yet routinely accomplished. But, if not a complete sequence, the use of antibodies can at least provide great detail as to the structure of the antigenic determinant sites and provides a reasonably simple method for resolving sequences of closely related proteins without the necessity of step-wise degradation studies.

It has been suggested that the optical properties of a polypeptide are the additive resultant of the properties of the individual constituent amino acids. The manner in which these optical properties relate to the elucidation of the specific sequence of small segments of peptides and to the secondary and tertiary structure of the total protein is the subject of the fifth chapter. In a similar light, the eighth chapter describes NMR which is useful in characterizing the nature of specific segments within the protein rather than in elucidating an entire sequence.

The final chapters deal with the most evolutionary procedures currently being applied to protein sequence determination: X-ray crystallography and mass spectrometry. Both are old techniques, each having seen extensive application to structural analysis of organic compounds. The list of proteins whose structures have been resolved by X-ray crystallography has grown rapidly over the past decade. For the most part, this has been accomplished in conjunction with the known chemically derived structure. However, with ever-increasing resolution, presently at about 1.5 - 2.0 Å, it becomes possible to complete the sequence analysis of a protein by X-ray diffraction procedures alone. The application of mass spectrometry to sequencing has not yet advanced as far as has X-ray crystallography. Nevertheless, it is entirely possible at this time to sequence relatively large peptides by mass spectrometry with no reference to ancillary procedures. The final chapter represents the first attempt to summarize the existing information on the application of NMR techniques to peptide sequence analysis.

As in the first volume, an attempt has been made to maintain the international character of presentations in the book. Distance presents its problems but, as before, the excellent cooperation of the recognized experts in each field is reflected in the successful attainment of this objective at the highest level of quality. It is with thanks and appreciation to those who were directly involved – and to many who made indirect contributions – that we bring this venture to its conclusion.

I have learned much from my masters,
more from my colleagues than from
my masters, and more from my pupils
than from all the others.

Judah the Prince
son of Rabbi Simon ben Gamaliel II
ca. 170 CE

Highland Park, Illinois
May 1977

SAUL B. NEEDLEMAN

Contents

Contributors

DOOLITTLE, RUSSELL F.	Department of Chemistry, University of California, San Diego, La Jolla, CA 92093, USA
FALTER, HERMANN	Department of Chemistry, Laurentian University, Sudbury, Ontario P3E 2C6, Canada
HORN, MARCUS J.	Department of Chemistry, Boston University, Boston, MA 02215, USA
KANNAN, K.K.	The Wallenberg Laboratory, University of Uppsala, S-751 22 Uppsala, Sweden
LAURSEN, RICHARD A.	Department of Chemistry, Boston University, Boston, MA 02215, USA
MROSS, GEORGE A.	Department of Chemistry, University of California, San Diego, La Jolla, CA 92093, USA
NEEDLEMAN, SAUL B.	Scientific Affairs, Abbott Laboratories, North Chicago, IL 60064, USA
NIEBOER, EVERT	Department of Chemistry, Laurentian University, Sudbury, Ontario P3E 2C6, Canada
REICHLIN, MORRIS	Department of Medicine and Biochemistry, Veterans Administration Hospital, State University of New York, Buffalo, NY 14215, USA

Step-wise Degradation of Peptides Attached to Solid Supports[1]

G. A. MROSS and R. F. DOOLITTLE

I. Introduction

The step-wise degradation of peptides was first successfully applied by Abderhalden and Brockmann (1) who used phenylisocyanate as an amino group coupling agent for the production of an intermediate which rearranged under acidic conditions, cleaving the terminal amino acid from the parent peptide. This procedure was extended by Edman (132), who changed the coupling agent to phenylisothiocyanate, a modification which yields a more readily cyclized intermediate than does phenylisocyanate. The method has been developed in a variety of modes (133) and reached a dramatic culmination in the form of an automatic sequencing device (134). Step-wise degradations have their limitations, however, and it is seldom possible to progress all the way down a peptide to the carboxy terminal amino acid when the chemical operations are conducted in free solution. The major problem involves the isolation of the cleaved residue from the parent peptide without significant loss of the latter. Ordinarily this is attempted by extracting the cyclic intermediate (thiazolinone) with an organic solvent in which the parent material is (hopefully) sparingly soluble. Unfortunately, depending on the amino acid composition of the peptide or protein under study, the losses are not insignificant and usually become greater as the peptide becomes smaller at each step. The attachment of peptides to insoluble supports via their α-carboxyl groups obviates that problem, while at the same time allowing the simple automation technology of the sort developed for solid phase peptide synthesis (298) to be employed. To this end, Laursen (244), in a pioneering effort, reported the attachment of a heptapeptide to polystyrene beads and its complete step-wise degradation.

During the past decade Laursen and his co-workers have introduced a number of innovations in this field as well as constructing a completely automated solid phase sequencer (249). Because these contributions are well reviewed elsewhere (250), in this article we will concentrate on procedures which have been primarily developed in and/or are currently in use in our own laboratory, while at the same time including some key recipes published by others. The procedures are presented with sufficient detail that any interested reader with a modicum of laboratory experience ought to be able to have solid phase sequencing under way as soon as the necessary reagents are obtained.

[1]Portions of this article are based on a Ph.D. dissertation submitted to the University of California, San Diego, by G.A. Mross.

II. Choice of Solid Support

Clearly the single most overriding feature in choosing a support is whether or not peptides can be attached in reasonable yield. Beyond that, different materials have certain mechanical properties which may be advantageous or disadvantageous, depending on the conditions to be employed. The major choice is between polystyrene beads on the one hand and glass beads on the other. Glass beads have the advantage of not shrinking or swelling in the presence of various organic solvents, but a considerable amount of early work in this field was accomplished with suitably derivatized polystyrene beads, and many of the orgininal published procedures deal with those materials.

A. Polystyrene Beads

Size and Cross-Linkage. In general, the smaller the bead, the more surface area available for derivatization and the more peptide one can hope to attach. Accordingly, 400 mesh is the bead size of choice when it is available, although 200-400 mesh is often used as a second choice. In a similar vein, the degree of divinyl cross-linkage ought to be as low as possible, which ordinarily means 1% (signified as X1 on most brand named resins, as in BioBeads-X1). A major supplier of resins suitable for peptide attachment has been Bio-Rad Laboratories (Richmond, Calif.)

Derivatization. The first step in the derivatization of polystyrene beads is chloromethylation. Suitably washed and dried beads are reacted with chloromethylmethyl ether and anhydrous stannic chloride at ice-bath temperature according to the procedure described by Merrifield (298). Chloromethylated resins also may be purchased from various manufacturers. The degree of substitution ought to be of the order of 0.7-1.5 mEq Cl^- per g of dry resin.

Chloromethylated resins can be reacted with a variety of substances to yield free amino groups; the most popular of which has been ethylene diamine to obtain aminoethylaminomethyl-polystyrene. The following procedure is quite straightforward; an alternative synthesis employing an intermediate nitration step leading to subsidiary ring amino groups has been described by Laursen (250).

To a slurry of 100 g of chloromethylated resin in dioxane is added 100 ml ethylenediamine. Stirring is maintained for four days at room temperature, after which the resin is washed with dioxane, 2N HCl, water, 2N NaOH, water and finally methanol. The resin is dried under vacuum and stored dry. Just before use, it is stirred with 2N HCl in 50% aqueous dioxane (10 ml/g resin), then filtered off and washed with methanol. The degree of amino substitution ought to be 0.2-0.6 mmol dry resin. Aminoethylaminomethyl-polystyrene beads may also be obtained from Bio-Rad Labs.

Other Amino-Substituted Polystyrene Supports. In addition to aminoethylaminomethylated polystyrene beads, simple amino polystyrene ("ring amino") and triethylenetetramine (TETA)[2] resin supports have also been employed

---

[2]Abbreviations used in this article include DITC, phenylene diisothiocyanate; DMF, dimethyl formamide; EDC, 1-ethyl-3-(3-dimethylaminopropyl)carbodiimide-HCl; TATG, thioacetylthioacetic acid; TEA, triethylamine; TETA, triethylenetetramine; TFA, trifluoroacetic acid; t-BOC, tert-butoxycarbonyl group.

A straightforward synthesis of the latter has been published by Horn and Laursen (197). Ten g of chloromethylated polystyrene (Bio-Rad S-X1, 400 mesh) are stirred with 125 ml of triethylenetetramine for 30 min at room temperature and then heated on a steam bath for 1.5 h. The resin is cooled, filtered and washed thoroughly with methanol. It is then stirred with 20 ml of triethylamine, filtered and washed again with methanol, then water, and finally methanol, before thorough drying under vacuum at 60°C. The TETA resin is reported to have both a higher capacity and reactivity.

B. Glass Beads

The success of workers using derivatized glass beads for immobilizing enzymes (462) led to their use as supports for solid phase degradation studies, and their successful deployment in this regard has already been reported (453). Glass beads are easy to handle in a mechanical sense, neither swelling nor shrinking as solvents are changed, and are readily derivatized. The following recipe for introducing amino groups is based on the initial studies of Weetal (462).

Aminopropyl Glass Beads. Ten g of Controlled Pore Glass (CPG-500) beads (Corning Glass) are washed thoroughly with 5N HCl, water and methanol and then dried under reduced pressure. The beads are then mixed with approximately 55 ml of 10% aminopropyltriethoxysilane (Union Carbide) in toluene (vol/vol) and maintained at 37° for 18 h with occasional stirring. The beads are then washed well with toluene, methanol, water and methanol again, before a final drying in vacuo. The number of amino groups per g of bead as determined by quantitative ninhydrin reaction is only slightly less than is obtained with polystyrene resins – to wit, about 0.1-0.2 nmol/g.

III. Attachment of Peptides to Supports

Certainly the most agonizing problem associated with solid phase degradation methods involves the attachment of peptides to supports in reasonable yields. As a rule, the degree of difficulty is roughly proportional to the size of the peptide and inversely proportional to the amount on hand. The attachment of peptides with fewer than ten residues seldom causes trouble if 100 nmol of material is available. Peptides with 10-50 residues are the ones which most investigators would like to affix, however, and the hope is often to start with 100 nmol or less and obtain attachment at something greater than 50% yield. Some of the following procedures are more successful than others in this regard, but there are a number of variables which are difficult to define, including the exact nature of the peptide under study, the type of support employed, previous investigatorial experience, etc. All of the following methods can be used with either polystyrene or glass beads after suitable derivatization. The coupling efficiency can be obtained by total acid hydrolysis of a weighed aliquot of the dried peptide-support product, and in the case of glass beads, ordinary constant boiling HCl (5.7 N) can be used in the same way as an ordinary hydrolysis (e.g., 24 h at 108°). For polystyrene beads, however, a 1:1 mixture of propionic acid and concentrated hydrochloric acid is more effective under the same conditions of time and temperature. If radioactive peptides are used, coupling efficiencies are obviously obtained simply by counting.

A. Carboxyl Activation Methods

The attachment of peptide α-carboxyl groups to amino groups on solid supports is clearly the most desirable mode of attack if the plan is to degrade the peptides from their amino-termini. Carboxyl groups – including but not exclusively the α-carboxyl – can be activated by a variety of agents to form intermediates which serve as targets for suitably nucleophilic amino groups on solid supports.

Activation:

Reaction:

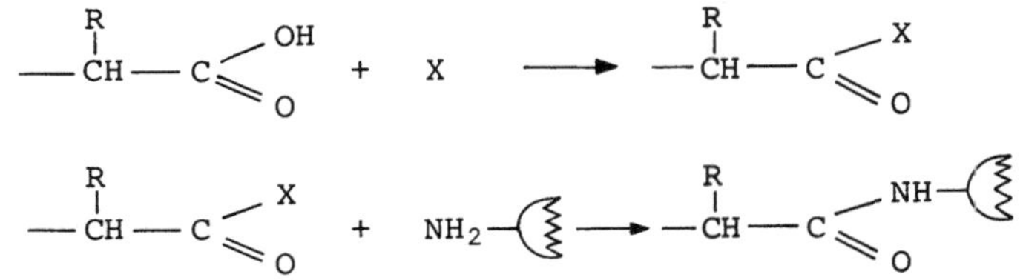

Protection of Peptide Amino Groups. It is desirable to block peptide amino groups in order to prevent both inter- and intra-molecular involvements with their activated carboxyl groups. Most peptides can be treated with t-butoxycarbonyl azide (t-BOC azide) in 50% aqueous dimethyl formamide (DMF) and all their amino groups blocked thereby. The reaction is carried out at room temperature for 18 h, although most amino groups are fully reacted within the first few hours (Fig. 1). Typically, 0.1-0.2 μmol of peptide are dissolved in 100 μl of 50% DMF, 10 μl of triethylamine (TEA) added and then 20 μl of t-BOC azide. After an overnight incubation, the solvent and excess reagents are evaporated in a desiccator over sulfuric acid and sodium hydroxide pellets. The t-BOC group is easily removed by exposure to anhydrous trifluoracetic acid (TFA) (e.g., 60 min at room temperature or 20 min at 40°) after coupling of the peptide to an amino-support. In the case of polystyrene supports it is advisable to use a 1:1 mixture of TFA and n-butyl chloride.

Carbonyl Diimidazole. The coupling agent originally used by Laursen (249) was carbonyl diimidazole. His recommended procedure involves activating the peptide by the addition of carbonyldiimidazole dissolved in DMF (40 mg/ml) and stirring for 1-2 h. Care must be taken to keep the system rigorously anhydrous. The amino-resin is then swollen with dry DMF (200 mg resin/ml DMF) for 30 min. Between one-third and one-half of the resin suspension is then added to the activated peptide. After

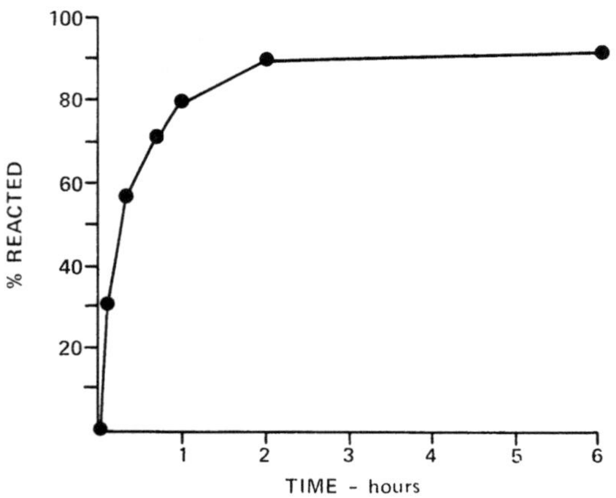

Fig. 1. Reaction of tert-butoxycarbonylazide (t-BOC-azide) with L-Lys-bradykinin. L-Lys-bradykinin (^{14}C-L-Pro$_{3,4}$) was reacted with 10% (vol/vol) t-BOC-azide in 50% aqueous DMF containing 5% triethylamine at 40°. At various time intervals, aliquots were removed and subjected to paper electrophoresis at pH 2 in order to separate unreacted, mono- and di-t-BOC peptides. Radioactivity was measured on a Packard Model 7201 Strip Counter, the areas under the peaks being used to calculate the peptide amounts. The percent di-t-BOC-peptide is shown in the graph. The radioactive Lys-bradykinin was the generous gift of Dr. J.J. Pisano

10 min, the remainder of the resin suspension is added and the incubation continued for 2 h. Unreacted resin amino groups are then blocked by the addition of an excess of methyl<u>iso</u>thiocyanate dissolved in 60% pyridine, pH 8-9. The resin is washed alternately with DMF and methanol several times and dried thoroughly.

Dicyclohexylcarbodiimide. N,N'-dicyclohexylcarbodiimide is the classic coupling reagent of peptide synthesis, and it can be used in the same mode to attach peptides to amino-supports (388). The peptide, with its amino groups blocked, is dissolved in 50-100 µl of a 10% solution (wt/vol) of dicyclohexylcarbodiimide in DMF, and the reaction carried out for 6 h at 37° (453). At that point the solution is mixed with the appropriate amino-support (suspended in DMF) and maintained at 37° for 18 h.

Water-Soluble Carbodiimides. Some of the water-soluble carbodiimides are more convenient to handle than dicyclohexylcarbodiimide in that they are powders and easily weighed. A typical protocol followed in our laboratory is the following: A t-BOC peptide (ca. 1 mg/ml in 50% aqueous ethanol) is added to dry amino-resin (ca. 1 g/mol peptide) and the solvent evaporated in a desiccator. The water-soluble carbodiimide, 1-ethyl-3-(3-dimethylaminopropyl)-carbodiimide-HCl (EDC), is made up to 0.1 M in 50% aqueous dioxane and the pH adjusted to between 4-6. Enough of the solution is added to the dry peptide-resin to just swell the resin (ca. 2 ml/g resin). The resin slurry is mixed well and left at 40° for 18 h, following which it is washed well with 50% dioxane, water and ethanol and dried.

Distinguishing Side-Chain Carboxyl Groups from α-Carboxyls. Ideally, peptide attachment ought to take place exclusively via the α-carboxyl group. In the methods described above, however, it is seldom possible to distinguish the α-carboxyl group from the side-chain carboxyls of aspartic and glutamic acids, and other strategies must be employed if those amino acids are to be identified during subsequent degradations. Two quite different approaches have been developed to solve this problem. First, side-chain carboxyls can be selectively blocked by treating a protein with some carboxyl-blocking agent in advance of fragmenting the protein into peptides, whether the fragmentation is via proteolytic enzymes or selective chemical breakage. In this case, only the carboxyl terminal peptide will be without a free α-carboxyl group, and all the other peptides derived from the protein will have exclusively unblocked α-carboxyl groups. The modification of protein carboxyl groups with glycinamide has been described by Hoare and Koshland (192); the method employs a water-soluble carbodiimide (EDC) as a coupling agent in the presence of 6 M guanidine chloride to unfold the protein. Alternatively, peptide carboxyl groups can be combined with glycinamide in the same way and the α-carboxyl block removed by the use of an appropriate enzyme. For example, in the case of tryptic peptides, trypsin will remove the glycinamide block from the carboxy termini of the derivatized peptides.

Oxazolinone Coupling. A much simpler way of distinguishing α-chain carboxyl groups from side-chain carboxyls has been formulated by Previero et al. (354). The method takes advantage of the tendency of carbodiimide-activated side-chain carboxyl groups to rearrange into unreactive urea derivatives. The corresponding urea derivatives do not form α-carboxyl carbodiimide complexes, an intermediate oxazolinone forming instead. As a consequence, if peptides containing both side-chain carboxyls and an α-carboxyl group are reacted with an appropriate carbodiimide, and the solution maintained for some length of time in the absence of any extra-molecular nucleophiles, the side-chain carboxyls

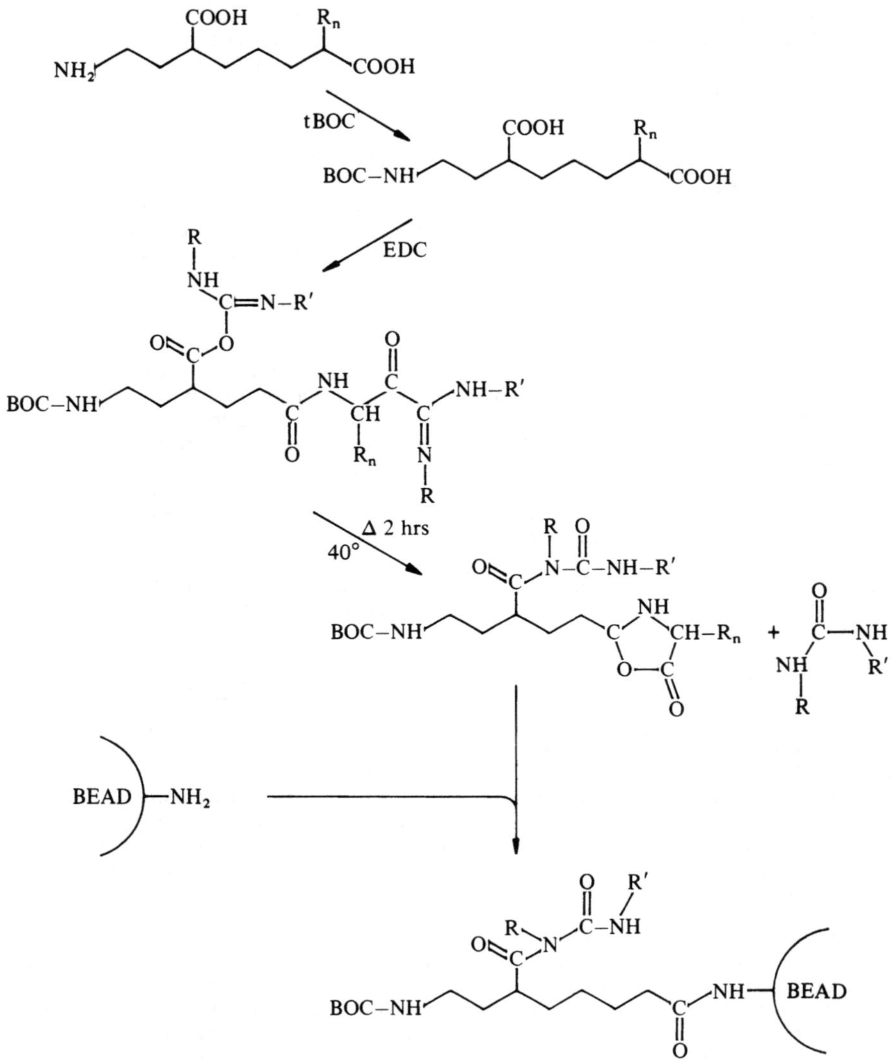

Fig. 2. Reaction scheme showing how peptides can be attached to amino-supports exclusively via their α-carboxyl groups. Adapted from Previero et al. (354)

rearrange into unreactive urea derivatives while the α-carboxyl group is maintained as a reactive oxazolinone. The postulated reaction scheme is depicted in Figure 2. The choice of carbodiimide is evidently very important, and the recommended recipe uses the same water-soluble carbodiimide (EDC) described above. It must be emphasized that in this case it is used in a nonaqueous environment, as outlined in the next paragraph.

The t-BOC peptide (50-250 nmol) is dissolved in 200 μl of DMF containing 1-2 μl of N-ethyl morpholine or TEA. Five mg of N-ethyl N'-(3-dimethylaminopropyl)carbodiimide (EDC) is added and the mixture let stand

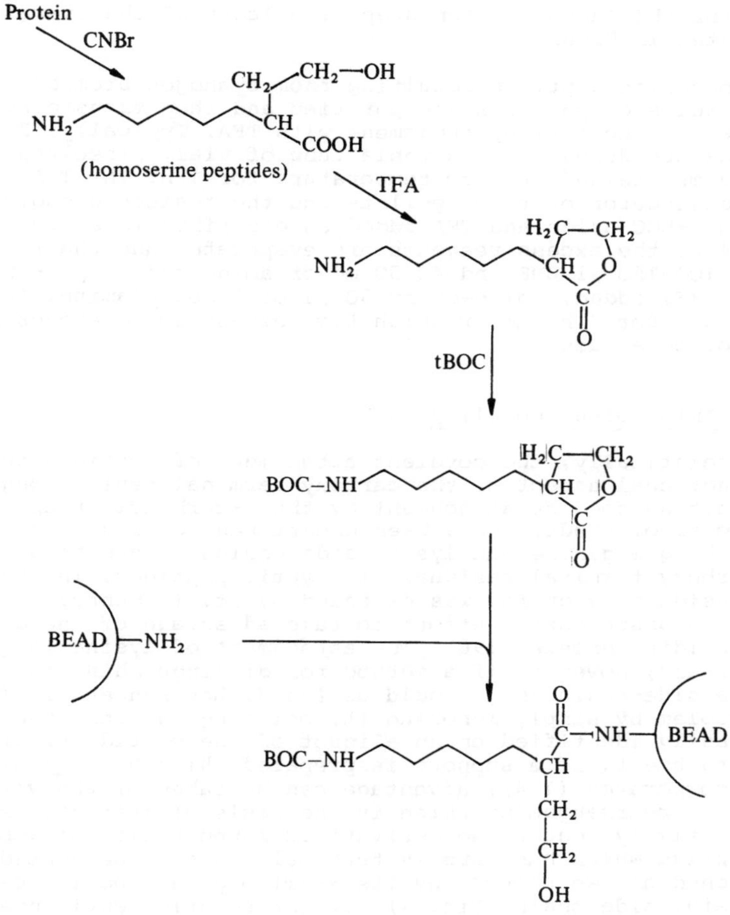

Fig. 3. Reaction scheme for attaching peptides with carboxy terminal homoserine residues to amino supports. Adapted from Horn and Laursen (197)

for 90 min at 40°. The amino-beads (50-150 mg) are then added and the system left for an additional 3 h at 40°. Excess amino groups on the support are then blocked with methylisothiocyanate as described above and the beads washed and dried (354).

Homoserine Peptides. Cyanogen bromide treatment of proteins or peptides results in a highly specific cleavage on the carboxyl side of methionyl residues, the methionine being converted to homoserine in the process (170). The relative sparsity of methionine in most proteins results in fragments of an average size well suited to solid phase sequencing. Accordingly, it was of great interest when Horn and Laursen (197) published a procedure for coupling peptides with carboxy terminal homoserine to amino-supports. The procedure depends on converting all of the homoserine residue to the homoserine lactone form, after which nucleophilic attack by the amino-resin is encouraged. Although they did not include a blocking step for peptide amino groups in their original recipe, that safeguard can be included if desired. It is important to note, however, that the peptide amino groups must be blocked after the lactonization step. The logic of the overall scheme is presented in Figure 3.

Homoserine peptides resulting from cyanogen bromide fragmentation of proteins or peptides are purified and then transformed completely to the lactone form by treatment with TFA. Typically, 20-200 nmol of peptide are dried in a suitable tube of vial, dissolved in 1 ml of TFA and maintained at room temperature for 1 h. The TFA is evaporated in a desiccator over KOH pellets and the residue dissolved in 100 µl DMF and t-BOC azide and TEA added as described in a previous section. After 2-4 h, the excess reagents are evaporated and the residue redissolved in 100-150 µl DMF and 40-50 mg of amino resin, pre-swollen in 200 µl of DMF, added, followed by 50 µl of triethylamine. The mixture is kept at 45° for 2 h, after which time excess amino groups on the resin are blocked as usual.

B. Amino Group Coupling

Theoretically, the covalent attachment of peptides to supports by any functional aspect of the carboxy terminal residue ought to be just about as good as attachment by the α-carboxyl group. Accordingly, a number of studies have been undertaken to find suitable ways of attaching arginine and lysine side-chains, since these are the common carboxy terminal residues of tryptic peptides. In the case of arginine, considerable energy was expended by Dr. T. Schroer in our laboratory in an unsuccessful effort to take advantage of the dione reactivity of guanidine groups (396). The attachment of lysine is quite straightforward, however, if a method for distinguishing the α-amino group from the side-chain amino could be found. Laursen et al. (248) solved this problem by simply ignoring it, pointing out that the α-amino acid is readily identified on an aliquot of the peptide treated without resin attachment. If a support is prepared which has isothiocyanate groups incorporated (124), advantage can be taken of the very same acid-catalyzed rearrangement which is the basis of most step-wise degradations. One simply reacts the derivatized support with an unblocked tryptic peptide which has carboxy terminal lysine. The peptide will now be attached at two places: by its α-amino group and its carboxy terminal lysine side-chain (Fig. 4). If the solution environment is now changed to favor cyclization of α-amino-substituted systems (by incubating with TFA, for example) then the α-amino residue will separate from the parent peptide even though both parts are still bound to the support. Cyclization cannot occur at the lysine side-chain, of course, so that site will stay bound throughout the degradation. All other residues, through to the penultimate one, ought to be removable as in an ordinary degradation (Fig. 4). Arginine peptides may also be attached by this procedure if the peptide is first deguanidinated by mild hydrazinolysis (248).

The conditions described by Laursen and his co-workers (248) are as follows. A peptide with carboxy terminal lysine is dissolved in 50 µl of N-methyl morpholine:water (1:1, adjusted to pH 9.5 with TFA). To this is added 1.0 mg of p-phenyldiisothiocyanate in 100 µl of DMF. The solution is heated at 45° for 30 min before transferring it to a second tube containing an amino-polystyrene resin (pre-swollen with DMF). The original tube is rinsed and the washing added to the resin mixture. After 75 min reaction at room temperature, any excess resin amino groups are blocked by the addition of methylisothiocyanate. After further incubation, the resin is washed and dried as usual before beginning a degradation series at the cyclization step.

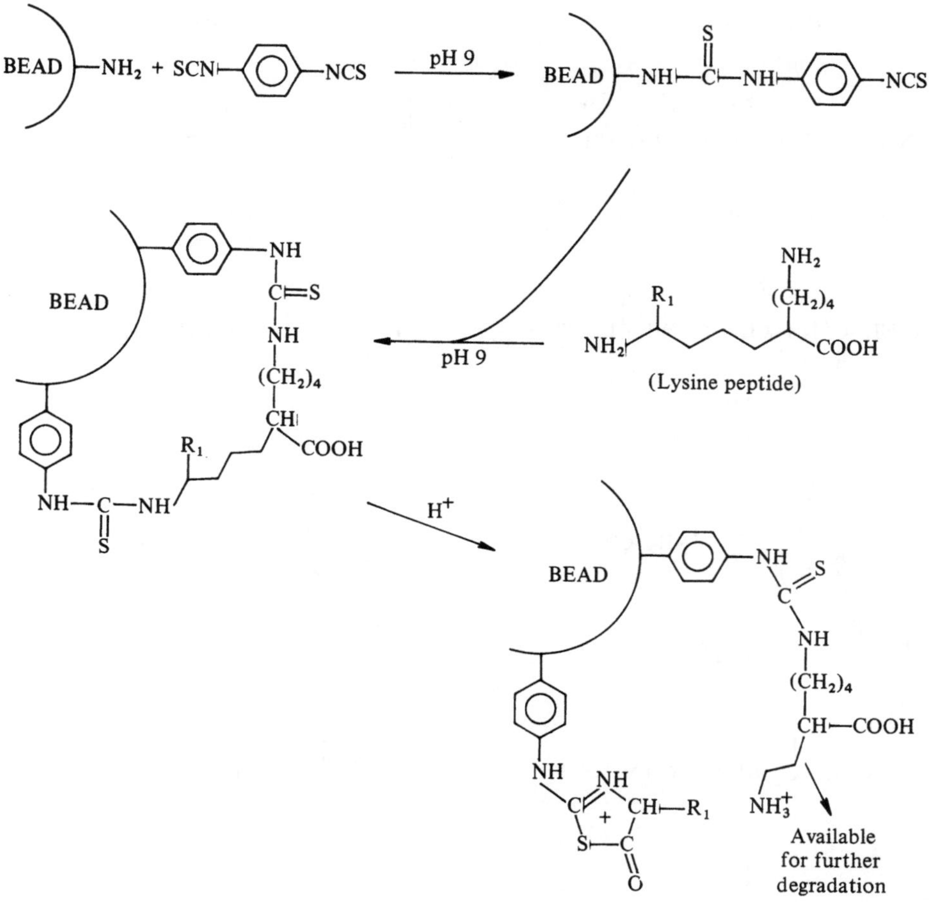

Fig. 4. Reaction scheme depicting the attachment of a lysine-containing peptide to an amino-support with phenyl diisothiocyanate. Based on the experiments of Laursen et al. (248)

The conversion of arginine side-chains to ornithine is accomplished by treating a suitable peptide with 0.5 ml 20% hydrazine at 105° for 30 min. This step may be done in an open tube equipped with a chimney or mini-still, or it can be performed in a sealed tube. In either case, the solution is subsequently taken to dryness on a rotary evaporator, following which traces of hydrazine are removed by co-evaporation of added 5% aqueous N-methyl morpholine. The peptide residue is then attached to the resin in the same way as described for lysine peptides.

Isothiocyanate Glass Beads. Wachter and colleagues (453) have treated aminopropyl glass beads with phenylene diisothiocyanate (DITC) to obtain the isothiocyanate derivative. The DITC reagent was used in a 25-fold excess to minimize cross-linking. The aminopropyl glass was added in small aliquots to a solution of DITC in dimethyl formamide (DMF). After the addition process, which took about 1 h, the mixture was allowed to stand for an additional 2 h, following which it was washed with DMF and the methanol, before drying in vacuo. The isothiocyanate beads were stored under nitrogen in a freezer until use.

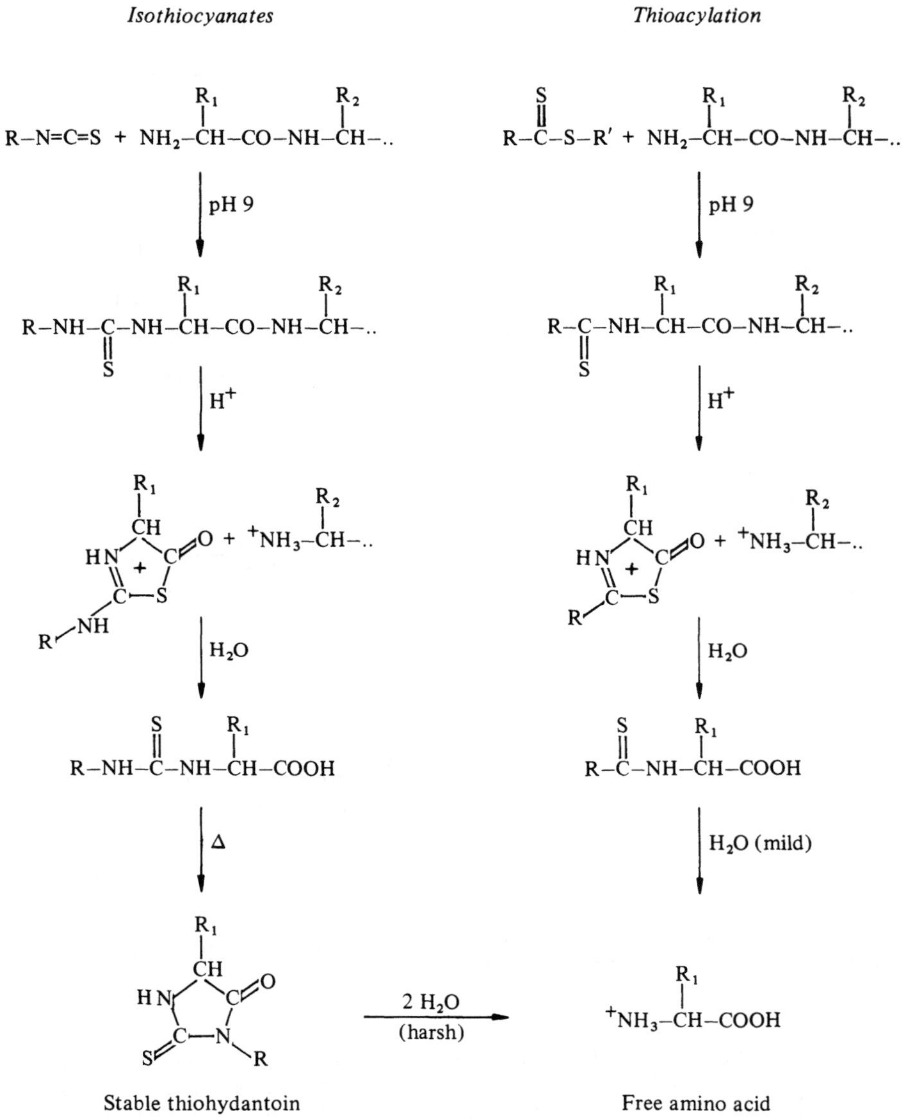

Fig. 5. Comparison of step-wise degradation pathways using isothiocyanates (left) and thioacylating agents (right). In the case of thioacylated systems there are no thiohydantoin intermediates, thus facilitating the regeneration of the free amino acid

IV. Step-wise Degradation Procedures

Once peptides are securely attached to a solid support, the investigator must decide on an amino-terminal degradation scheme. The choice ought to be made consistent with the facilities available for identifying the products obtained at each degradative cycle. If an isothiocyanate reagent is used, the product will ordinarily be a thiohydantoin, and these may be identified by paper or thin layer chromatography or by more sophisticated gas chromatographic procedures or high-pressure liquid chromatography. It is also possible to back-hydrolyze the derivatives to the free amino acids (416), but the conditions are very harsh and somewhat inconclusive. Thioacylation degradative procedures, on the other hand, do not produce thiohydantoins, being restricted to a thiazolinone intermediate, and as a result return the free amino acid after relatively mild hydrolysis (Fig. 5). It is also possible to identify the thioacylated amino acids directly.

A. Isothiocyanate Methods

The most popular step-wise degradation procedure in current usage is that described by Edman (132) employing phenylisothiocyanate, and a complete description of a manual form of this procedure has been described in a previous volume of this series (133). In general, the same recipes can be used for solid phase work if appropriate solvent substitutions are used with certain supports. For example, polystyrene beads float in dichlorethane, but n-butyl chloride substitutes perfectly well. On occasion (244), methylisothiocyanate has been used instead of phenylisothiocyanate, the methyl hydantoins having certain advantages in some identification systems. Either methylisothiocyanate or phenylisothiocyanate can be obtained as their ^{35}S-derivatives, the radioactive reagents improving the sensitivity (and often the quantitation) of the procedures.

B. Thioacetylation Method

The possibility of using N-thioacyl peptides for sequence analysis as an alternative to isothiocyanate degradation dates back to Kenner and Khorana (233), who employed the somewhat similar dialkyl xanthates, and who noted the simplicity with which the free amino acids could be regenerated. Previously Kjaer (236) had reported the thiobenzoylation of amino acids and their amides, and, more recently, Barrett (32) and then Previero and Pechere (353) showed that the N-thiobenzoyl group has a labilizing effect on amino-terminal residues comparable to that of phenylthiocarbamylation. In both cases, S-alkyldithiobenzoates were employed to thiobenzoylate the peptide and anhydrous TFA used for cleavage. During the course of exploring other thioacylating possibilities (310), we synthesized a number of di-alkyldithioesters according to published recipes. Of all of these, thioacetylthioglycolic acid (TATG) has proved the most satisfactory for solid phase usage and is the reagent with which all the following studies were conducted (309). We have also employed the related compound thioacetylthioethane (TATE), especially for peptide degradations in free solution. Either reagent can also be used to good advantage in the study of proteins not attached to supports (424).

Synthesis of TATG. The reagent may be prepared by either of two published procedures. Jensen and Pedersen (216) started with N-acetyl piperidine which they converted to the thioester by refluxing with phosphorus pentasulfide in pyridine. The thioacetyl piperidine is alkylated with bromoacetic acid in benzene and then thiohydrolyzed with H_2S in ethanol to yield the product. A simpler route is based on a recipe for a similar substance described by Marvel et al. (283) and is the procedure we now recommend. Acetonitrile is reacted with mercaptoacetic acid to yield carboxymethylthioimidate which is then converted to the product by treatment with H_2S in pyridine at $0^{o}C$.

$$CH_3CN + HS\text{-}CH_2COOH \longrightarrow CH_3\overset{\overset{NH}{\|}}{C}\text{-}S\text{-}CH_2COOH \xrightarrow{H_2S} CH_3\overset{\overset{S}{\|}}{C}\text{-}S\text{-}CH_2COOH$$

The yields of the reaction are good, and 10 g lots of the yellow solid can be prepared in short order by any competent undergraduate who has had a course in organic chemistry.

TATG Recipe. Mercaptoacetic acid (46 g) and acetonitrile (22.5 g) are mixed in a reaction flask and cooled to ice temperature. The mixture is overlaid with 2-3 cms of petroleum ether and HCl gas bubbled through the lower phase until a white solid appears (carboxymethylthioimidate). The solvent is then flash evaporated and 250 ml of dry pyridine saturated with H_2S added. Hydrogen sulfide is then bubbled through for 6 h or until the reaction is complete. During that time the white slurried solid will liquify and turn yellow, following which a white precipitate (ammonium chloride) will appear. When the reaction is completed, add 100 ml ice water with stirring and let stand 5 min to dissolve the ammonium chloride. The liquid is then poured over a mixture of 300 ml concentrated HCl and 100 ml H_2O and 300 g ice. The system is extracted with 300 ml diethyl ether. After separation of the phases, the aqueous layer is extracted twice more with 150 ml ether each time. The combined ether layers are washed with 3N HCl, following which the ether layer is dried with sodium sulfate. The ether is evaporated (rotary evaporator), yielding about 20-25 g of oil. TATG can be crystallized from warm carbon tetrachloride; the yield is 10-12 g of yellow solid (m.p. = 80-81°C).

Efficiency of Coupling and Cleavage. The two key steps in any step-wise degradation are coupling and cleavage, both of which must occur at very high efficiencies if an extended degradation is to be successful. When TATG is used as a coupling agent at pH 9 in 50% pyridine, coupling is virtually complete within 20 min (Fig. 6). In manual methods we generally let the incubation continue for an hour; in our automated sequencer we employ two separate 30-min couplings to insure maximal thioacetylation. As noted by Barrett (32), cleavage of thioacylated with

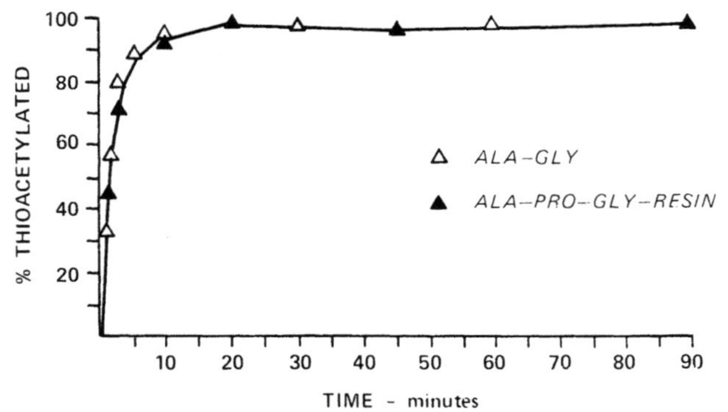

Fig. 6. Rates of thioacetylation of L-alanyl-glycine (in solution) and L-alanyl-L-prolyl-glycine (on a polystyrene resin). The Ala-Gly preparation (2 mM, △) was reacted with TATG (0.33 M) in aqueous solution at pH 9.0 at 40°. The percent alanine coupled was determined by measuring the unreacted dipeptide (unhydrolyzed) on a standard amino acid analyzer. The Ala-Pro-Gly (0.16 µmol, ▲) was reacted with 0.3 M TATG in 60% pyridine, pH 9.2, 40°. In this case, the percent coupled was determined not only by the amount of alanine released after cleavage by TFA but also by the amounts of alanine remaining on the thoroughly washed resin

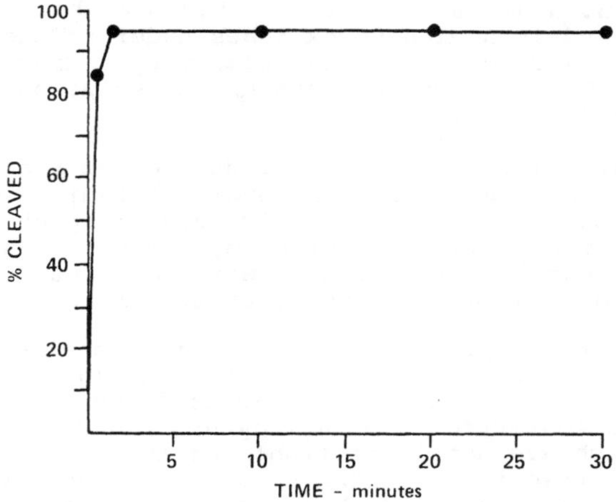

Fig. 7. Rate of cleavage of the amino-terminal residue from thioacetylated Ala-Pro-Gly attached to a polystyrene resin. The thioacetylated peptide-resin (0.12 μmol in each of a series of tubes) was suspended in a 1:1 mixture of TFA:dichlorethane at 40°. At intervals, one of the reaction mixtures was terminated by the addition of cold pyridine/triethylamine (9:1 by vol) and the resin washed and dried. The resin was then acid-hydrolyzed and the amounts of uncleaved alanine determined

TFA is also a very rapid process. In our system cleavage is maximal in less than 5 min when conducted at 40° (Fig. 7).

Extracting the Thiazolinone. A wide variety of reagents can be used for extracting the thiazolinone since there is no danger of removing the parent peptide, which is securely attached to the solid support. In the first place, the TFA itself may be harvested without added extractant, and this is the method of choice in an automated system. Otherwise, ether, butyl chloride, dichlorethane or ethyl acetate – and doubtless other solvents – can be used. In any event, it is important to mix the extract with a small amount of water or 5.7N HCl *before* evaporating the solvent, since some of the 2-methylthiazolinone amino acids are volatile enough to be lost. The addition of a little water or aqueous acid rapidly converts the thiazolinone to the thioacetylated amino acid which, after the removal of the volatile organic solvent, can be hydrolyzed to yield the free amino acid.

Regeneration of Amino Acids. Most of the thioacetylated amino acids are readily converted to the free amino acid by acid hydrolysis with 300 μl

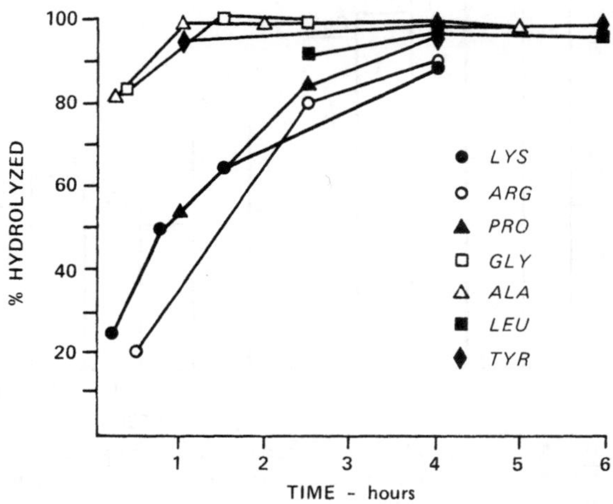

Fig. 8. Hydrolysis rates of various amino acid 2-methylthioazolinones derived from thioacetylated peptides. Hydrolysis was accomplished with 0.2 ml 5.7N HCl at 120° in sealed tubes. The hydrolysates were dried down in a desiccator over NaOH and analyzed on a Spinco Amino Acid Analyzer. The 100% level was established by hydrolysis and analysis of corresponding amounts of the original peptide in each case

of 5.7N HCl at 125° for 16 h. In either case, the hydrolyses are performed in sealed Pyrex tubes under reduced pressure. The amino acids can also be regenerated simply by adding an equal volume of water to the TFA extract and heating the solution in a sealed tube as described for 5.7N HCl.

Some amino acids are lost during the hydrolysis step, including tryptophan and threonine. Moreover, glutamine and asparagine are converted to glutamic acid and aspartic acid, respectively. The identification of tryptophan is not a problem, however, since the extracted derivative is usually quite orange. Moreover, hydrolysis with methane sulfonic acid (266) gives a quantitative yield of free tryptophan.

Threonine presents a more serious challenge, however. Both threonine and serine undergo dehydration during the TFA cleavage step as a result of a β-elimination. During acid hydrolysis the unsaturated derivatives are converted to methyl pyruvate and pyruvate, respectively, compounds which are not identifiable on an amino acid analyzer. If mercaptoethanol is present in the TFA cleavage system, however, the mercaptan readily adds across the double bond of the serine derivative, and subsequent acid hydrolysis yields hydroxyethylcysteine in good yield, a compound which fortuitously emerges from the analyzer at the same position as threonine. Unfortunately, the corresponding thiol addition does not occur in the case of the unsaturated threonine equivalent, and a suitable alternative is still being sought. The thioacetyl derivative can be identified, however, and it may prove necessary to remove an aliquot of extractant prior to hydrolysis at every step.

Glutamine and asparagine can be distinguished from glutamic acid and aspartic acid if the side-chain carboxyl groups are blocked with glycinamide (Sect. III-A), in which case an equivalent amount of glycine appears in the analyses of glutamic acid or aspartic acid. Even if side-chain carboxyl groups are not blocked, the free acids can usually be distinguished from glutamine or asparagine if a carbodiimide coupling is used, since invariably some of the side-chain carboxyls also couple and reduce the yield (Fi.g 9).

Even though one of our long-term objectives has been the identification of the regenerated amino acids, it may prove profitable to develop a

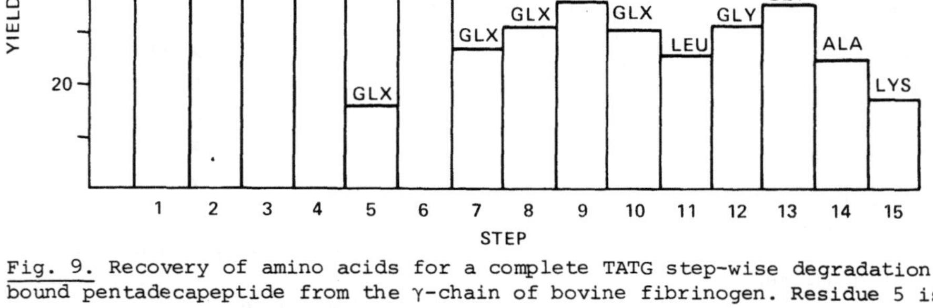

Fig. 9. Recovery of amino acids for a complete TATG step-wise degradation of a resin-bound pentadecapeptide from the γ-chain of bovine fibrinogen. Residue 5 is actually glutamic acid, whereas the residues 7, 8 and 10 are glutamines (88). Performed manually

high-pressure liquid chromatography system for the identification of the thioacetylated amino acids. In addition to eliminating the acid hydrolysis step, identification at this stage would also distinguish glutamine and asparagine from glutamic acid and aspartic acid as well as providing a quantitative measure of tryptophan and threonine.

Details of the TATG Manual Procedure. Approximately 50-200 mg of glass (or polystyrene) beads with 50-200 nmol of peptide covalently attached are placed in a small glass-stoppered tube (vol. ca. 3-5 ml). If the peptide has not yet been de-blocked (i.e., removal of BOC group), then the beads are covered with 200-300 μl of TFA and incubated at 40° for 20 min, following which the TFA is evaporated in a desiccator containing KOH pellets. The beads are then washed twice with 2 ml of methanol, partially dried by aspiration and then dried further under vacuum.

Coupling. The coupling reagent consists of 20 mg TATG/ml 50% pyridine. It is neutralized by the addition of TEA (20 μl/ml), more being added, if necessary, to raise the pH to 9.0-9.5. Usually 0.5 ml of the solution is added to the beads, the mixture gently agitated and placed at 40° for 60 min. After coupling is complete, the beads are washed twice with 50% pyridine and three times with methanol before drying. The washing procedure is simple, involving addition of the solvent, agitation on a Vortex agitator, centrifuging for 30 sec in a table-top centrifuge and removal of the solvent with a disposable Pasteur pipette. A series of five washes can usually be accomplished in 5-10 min.

After the beads are thoroughly dried (usually 1-16 h in a vacuum desiccator), the cleavage step is conducted by the addition of 200 μl of TFA containing 5 μl/ml mercaptoethanol. The system is placed at 40° for 15 min, following which the cleaved residue is extracted with two 1-ml extractions of diethyl ether (or n-butyl chloride, ethyl acetate, etc.). The extracts are added directly to 7 mm (I.D.) Pyrex hydrolysis tubes containing 100 μl of 5.7N HCL, and the system carefully mixed before evaporation of the solvent. After the organic layer has been removed in a stream of nitrogen, an additional 200 μl of 5.7N HCl is added, the hydrolysis tube evacuated with a vacuum pump and sealed. The solution is placed at 125° for 4 h (or 105° for 16 h), after which the tubes are opened, and the acid evaporated in a desiccator over KOH pellets. The residues are dissolved in pH 2.2 buffer and suitable aliquots applied to an ordinary amino acid analyzer. In the meantime, the beads are washed with methanol twice and recycled through the procedure. As in most degradations, the beads themselves never leave the original reaction tube.

V. Automation of Solid Phase Degradation

In actuality, solid phase degradations are considerably simpler to automate than are the solid phase syntheses upon which they are patterned. This is because the operations are always exactly the same in the case of the degradations, whereas in the case of synthesis the system must be programmed to incorporate a different one of twenty amino acids at each cycle. The degradative steps are simply coupling, washing and drying ("laundering"), cleavage, extraction, a little washing, and back to coupling again. Only five or six reagent reservoirs are necessary for all operations.

One choice to be made in the automation of a solid phase degradation device is the kind of reaction vessel. Laursen (249) employs a glass column and packs the peptide beads in with (underivatized) glass beads;

the various solvents are pumped in and out. We use a chamber similar to those used in solid phase synthesis, except jacketed. The reaction vessel consists of a 5-ml jacketed tube fitted at the bottom with a sintered glass frit of sufficiently small pore size, so that it is not readily clogged by beads. The lower end of the vessel is a glass male Luer fitting which connects with the inlet-outlet line. At the top is a female ground glass joint into which fits the ground glass vessel cap. The cap also contains a sintered glass frit, and it also extends into a glass male Luer fitting for connection with a nitrogen (and/or aspirator) line.

A. Semi-Automated Operation

We were able to fashion a very simple semi-automated solid phase sequencer from a commercially available peptide synthesizing unit (Chromatronix Peptide Synthesizer Switcher). Basically, the system consists of three Kel-F-coated 6-way rotary valves which are readily plumbed in a way to select the various reagents, nitrogen inlet, water aspirator for removal of solvents (optional), and a selector for diverting unwanted washings down the drain or collecting the extracted thiazolinone in a tube (Fig. 10). The solvents are pushed into the chamber by nitrogen pressure. The reaction vessel is mounted on a shaker which rocks over approximately 180°. A water bath is connected to the jacketed vessel in order to perform all the operations at 40°. The operations are very simple, and a complete cycle can be conducted in 85-100 min (Table 1). Four or five degradation cycles are easily completed in an average working day. Indeed the octapeptide depicted in Figure 11 was completely sequenced in two working days using this device.

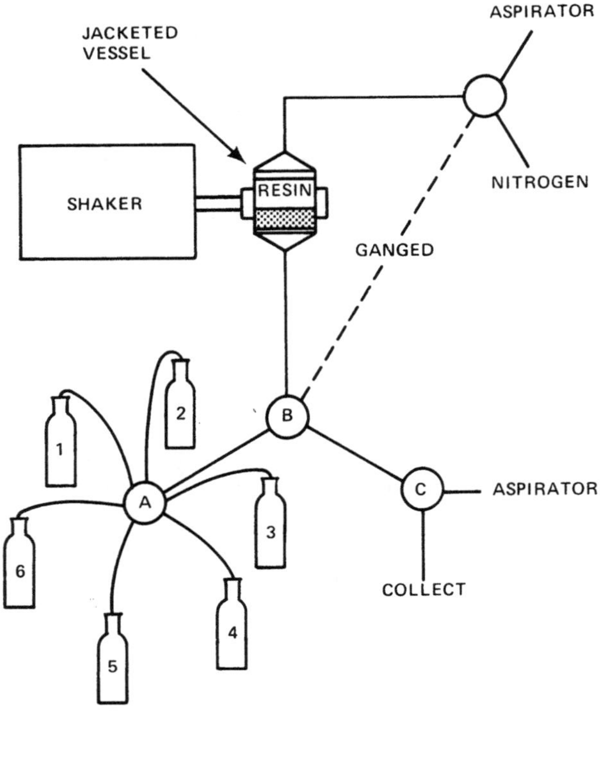

Fig. 10. Schematic outline of a simple semi-automated apparatus for the step-wise degradation of peptides attached tc solid supports. Vessels 1-6 contain all the necessary solvents (coupling reagent, water, 50% pyidine, methanol, TFA and butyl chloride). A, B, and C are Kel F-coated 6-way rotary valves. The system is easily adapted from commercial peptide synthesis switching devices

Table 1. Semi-automated TATG degradation of peptides attached to solid supports. List of operations and approximate duration of each

Operation	Duration (min)
1. Equilibration with coupling buffer[a]	5
2. Thioacetylation with TATG in coupling buffer	45-60
3. Wash with coupling buffer (2-3 times)	5
4. Wash with methanol (2-3 times)	5
5. Drying with nitrogen	10
6. Cleavage with TFA (collect)	10
7. Wash with methanol	5
	Time elapsed 85-100 min

[a]Unless otherwise noted, all operations are performed one time per cycle. When the automatic system is used, coupling and cleavage sequence are executed twice each for maximum efficiency.

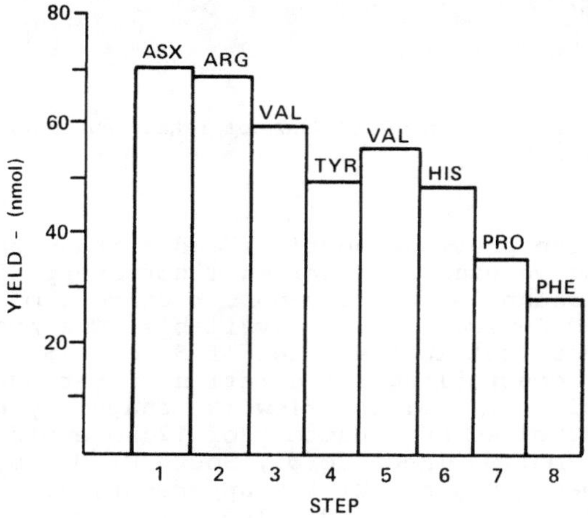

Fig. 11. Recovery of amino acids for a complete TATG step-wise degradation of Asn_1-Val_5-angiotensin II coupled to an aminoethylaminomethyl polystyrene resin. Performed with the semi-automated arrangement. The peptide was the generous gift of Dr. J.J. Pisano

B. An Inexpensive Automated Apparatus

As noted above, Laursen (249) has published the complete specifications of an automated sequence device.[3] What follows here is a brief description of a much less sophisticated apparatus which can be assembled for a very modest cost.

Basically, the unit consists of a programmer for switching on solenoid valves in a specified sequence, an array of 12 solenoids for controlling solvent inlet and outgo, a water-jacketed reaction vessel mounted on a shaker (the same one used in the semi-automated apparatus described

[3]Available from Sequomat, Inc., Watertown, Mass.

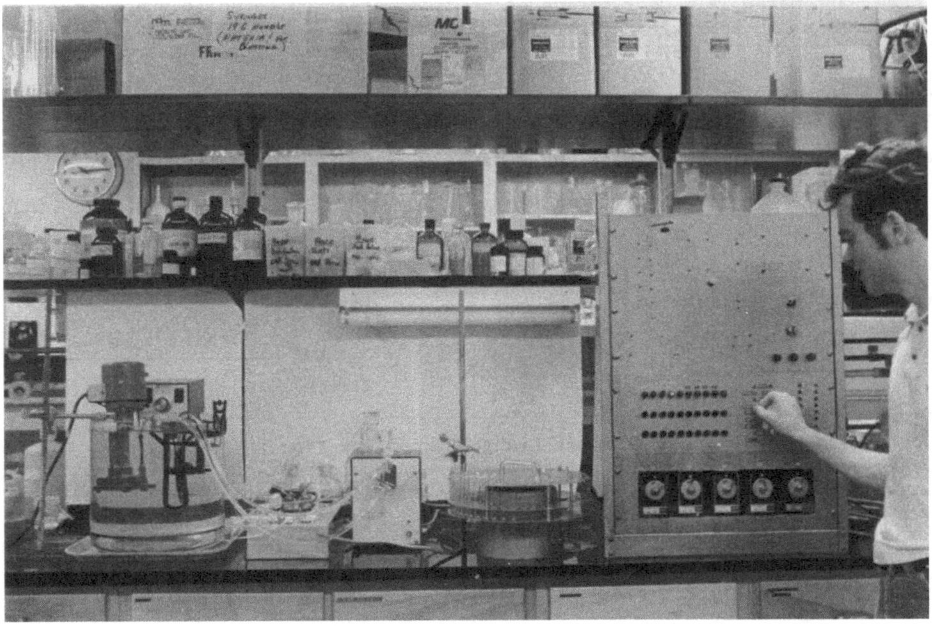

Fig. 12. Photograph of the completely automatic solid phase sequencer outlined in Figure 13

in the previous section) and a fraction collector (Fig. 12). In our case we used a discarded Tenor Stepping Drum Programmer (100 steps, 31 channels) but any number of more modern and less expensive progamming devices are now available. The programmer is equipped with five times (set in the range of 5 sec, 15 sec, 50 sec, 5 min, and 15 min) for controlling the duration of various operations in the degradative cycle. All solvent flow is managed by nitrogen pressure (2 psi) and controlled by a battery of 12 solenoid valves (Fig. 13). The solenoid assemblies (Mace Corp., South El Monte, Calif.), which are the most expensive parts of the apparatus, are judiciously positioned to minimize dead space and solvent intermixing. In this regard, the TFA reservoir is guarded by valves on the nitrogen side as well as on the output side. A complete description of the apparatus has appeared elsewhere (118).

Automation of solid phase degradation methods is not only labor-saving, but it also can be more efficient in a chemical sense. For example, each cycle can be programmed to include double-coupling and double-cleavage routines, thereby insuring virtual completeness at these two critical steps. Finally, it must be noted that the solid phase sequencer can be used for any degradative scheme simply by changing the reservoir contents.

VI. Concluding Remarks

When we began these studies some years ago, we had two objectives in mind. First, we wanted to be able to go all the way down to the end of

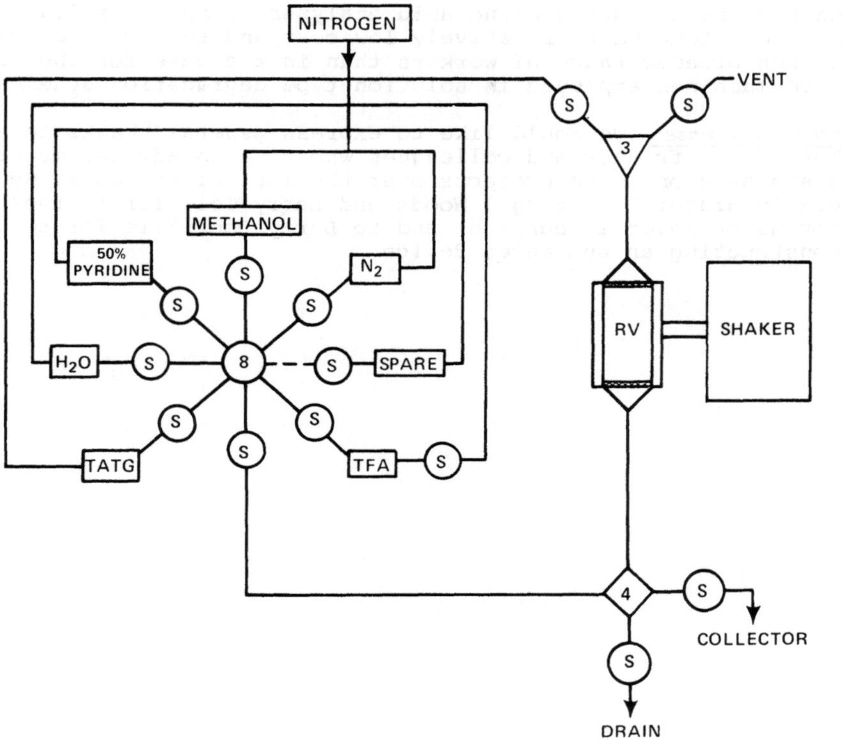

Fig. 13. Schematic outline of a completely automatic solid phase sequencer designed for use with the TATG thioacetylation method. All solvents are pumped by nitrogen pressure (ca. 2 psi) and controlled by a programmer which activates five timers, twelve solenoids, the shaker and the fraction collector (118)

peptides during step-wise degradation; second, we wanted a quantitative identification method which was unambiguous. The first of these objectives was met by following Laursen's lead and the use of covalently bound solid supports. The second was attained by the use of a thioacylating procedure which regenerates the free amino acid for identification on a standard amino acid analyzer.

In this chapter we have tried to point out the advantages of solid phase peptide sequencing as well as some of its shortcomings. High yield attachment of peptides to suitable solid supports is still the biggest hurdle, although a number of very promising methods are now available. Of these, the two novel procedures introduced by Laursen and co-workers (197, 354) will probably be the most useful in the long run, since the homoserine lactone attachment method can be used in combination with the amino-coupling procedure to obtain all the overlapping peptides of a given protein. Once the peptides are covalently attached to the beads, the advantage is won. At this point, the majority of workers will probably choose to employ a conventional isothiocyanate degradation. On the other hand, in the paragraphs above we have described a thioacetylation procedure which has the advantage of unequivocal identification of the released residues. After years of torment involving vague spots on thin layer or paper chromatograms, we are delighted with the unambiguous certainty of answers returned on a Moore and Stein amino acid analyzer. Finally, solid phase methods can be automated at relatively low cost and thus can be available to a much broader range of workers than is the case for the very expensive machines employed in solution-type degradation schemes.

Acknowledgments. We would like to express general thanks to a large number of our friends and colleagues who gave us advice, materials and assistance on these projects over the last eight years. We are especially grateful to Virgil Woods and Larry Toll for synthesizing TATG for us on several occasions and to Larry Doolittle for designing and constructing an automated device.

Coupling Methods and Strategies in Solid-Phase Sequencing

R. A. LAURSEN and M. J. HORN

I. Introduction

The success of solid-phase sequencing depends on the availability of reliable procedures for immobilizing peptides and proteins. Once the peptide is attached to an insoluble support, sequencing generally proceeds without problems, provided the peptide has an unblocked α-amino group. Since the introduction of the solid-phase Edman degradation in 1966 (244), considerable effort has gone into the search for a completely general procedure for immobilizing peptides. The search has thus far been unsuccessful, but in the process, several attachment methods that take advantage of the reactivity of specific amino acids have been discovered. This multiplicity of immobilization procedures, rather than limiting the utility of the solid-phase Edman degradation, has given it a versatility that was unanticipated.

This chapter will discuss the pros and cons of currently available peptide coupling methods and how they can be used to maximum advantage in the sequence analysis of proteins. Detailed procedures and instrumentation are discussed in Chapter 1 and in recent reviews (245-247).

II. Coupling Procedures

A. Attachment by Activation of Carboxyl Groups

1. *Carbonyldiimidazole*

The first carboxyl activating agent to be used in conjunction with solid-phase Edman degradation was carbonyldiimidazole, which was introduced to peptide chemistry by Paul and Anderson (338) in 1960. It was chosen because of its reported (338) high yields and absence of side reactions. Activation occurs by way of the imidazolide (Fig. 1).

Carbonyldiimidazole reacts rapidly with nucleophiles such as amines and water. For this reason the reaction must be carried out under anhydrous conditions in a solvent such as dimethylformamide, which has

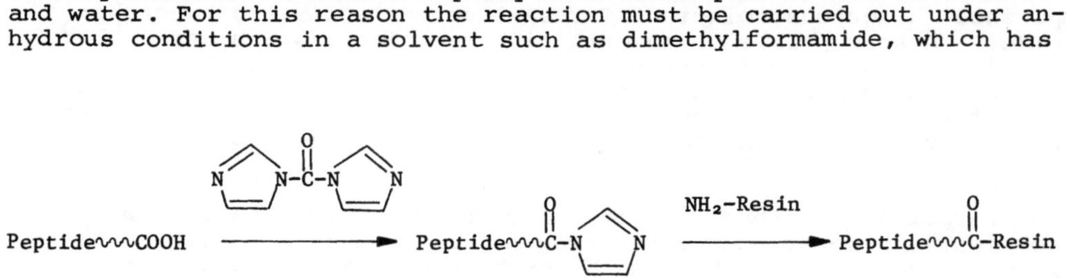

Fig. 1. Carbonyldiimidazole activation

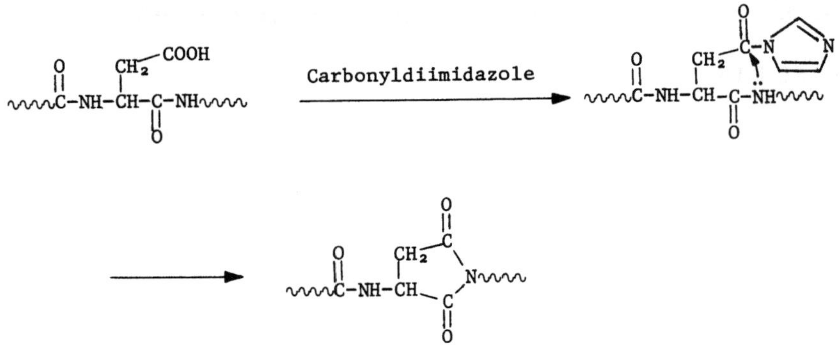

Fig. 2. Cyclic imide formation

been distilled from P_2O_5 to remove traces of water and dimethylamine. The peptide amino groups must be protected, for example, with the *t*-butyloxycarbonyl group (249), which is easily removed under the conditions of the Edman degradation.

Activation of the carboxyls is usually done with a several hundred-fold excess carbonyldiimidazole, which should be destroyed after activation to prevent reaction and cross-linkage of resin amino groups. This is easily accomplished by adding an approximately stoichiometric amount of water to the activation mixture. Under these conditions the carbonyldiimidazole is destroyed in about 10 min, whereas the peptide imidazolide is relatively stable (245, 246).

Although the carbonyldiimidazole method gives relatively high attachment yields, it suffers from several disadvantages. First of all, not all peptides, especially larger ones, are soluble in anhydrous dimethylformamide. Second, one must exercise special care to be sure that anhydrous conditions are maintained. The most serious problem is that side-chain carboxyls as well as the C-terminal carboxyl, are activated. In the case of glutamic acid, the α-carboxyl becomes attached to the resin, with the result that a blank appears in the sequence at Glu, because the thiazolinone remains attached to the resin. In the case of aspartic acid, degradation stops at Asp, presumably because of cyclic imide formation, which prevents further degradation (Fig. 2).

2. Carbodiimides

Carbodiimides have long been used in peptide synthesis to form amide bonds, and have the advantage over carbonyldiimidazole that they are relatively insensitive to water. Thus peptide attachments can be accomplished in aqueous solution. Carboxyl groups activated by carbodiimides are susceptible to several side reactions, which may either be a help or hindrance in efficient coupling of peptides (Fig. 3). The reaction of a carboxyl with a carbodiimide gives first an o-acylurea which, if an amine is present, reacts to form an amide. Competing with this reaction is rearrangement to an N-acylurea, which is no longer reactive with nucleophiles.

If the carboxyl is C-terminal, an azlactone can form, which either hydrolyzes with water to the starting material or reacts with an amine to form an amide bond. Activated side-chain carboxyls either react with amines to form amides or rearrange to the inactive N-acylurea.

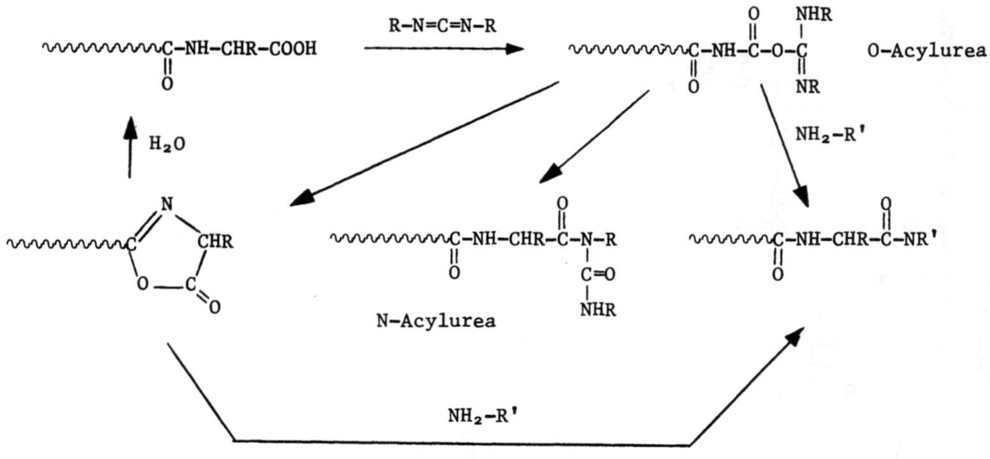

Fig. 3. Reactions of carbodiimide-activated carboxyl groups

In the case of Asp-Gly bonds, some cyclic imide may be formed as noted with carbonyldiimidazole.

A variety of carbodiimide coupling procedures has been described. The simplest involves adding an excess of N-dimethylaminopropyl-N'-ethyl carbodiimide to a mixture of the unblocked peptide and resin in dimethylformamide-water (246, 389). In this case the great excess of resin amino groups over peptide amino groups apparently prevents polymerization of the peptide. All carboxyls become linked to the resin. Alternatively, one can block the amino groups with *t*-butyloxycarbonyl azide (245, 246).

The major problem – that of selective blocking of the side chain carboxyls – has been approached in two ways. In the case of tryptic peptides, Mross and Doolittle (see Chap. 1) have converted all the carboxyls to amides by reaction with glycinamide and a carbodiimide, and then selectively released the C-terminal by treatment with trypsin. This procedure requires a separate step to separate the peptide from excess glycinamide. Previero et al. (354) activated peptides with a water-soluble carbodiimide for about 90 min in the absence of a nucleophile. During this time the O-acylurea intermediate of the side-chain carboxyl can isomerize to the inactive N-acylurea (Fig. 4). However, the environment at the C-terminal carboxyl is unique in that azlactone formation is possible. Addition of the amino resin to the azlactone results in a peptide bound specifically at the C-terminus. Unfortunately the yields of peptide linkage by this and other versions of the carbodiimide attachment procedure are frequently very low and unpredictable. The most likely reason for the low yields is N-acylurea formation at the C-terminus, which competes with the desired reactions and inactivates the peptide.

3. Homoserine Lactone

Cleavage of proteins at methionine with cyanogen bromide (171) generates peptides having C-terminal homoserine residues. Treatment of the peptide with trifluoroacetic acid converts homoserine to the lactone (11), which is then sufficiently activated to react with nucleophiles

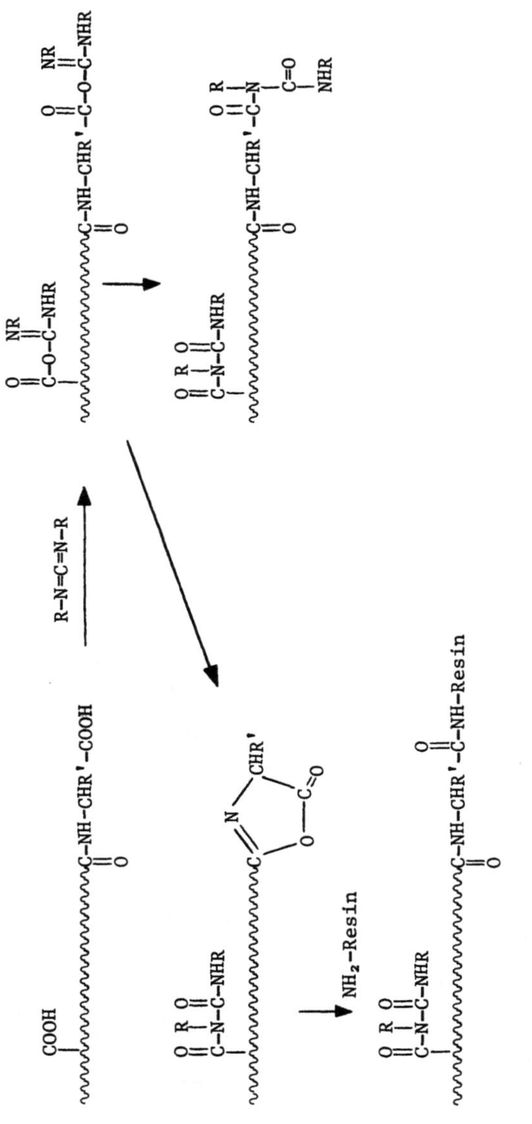

Fig. 4. Selective activation of peptide C-terminal carboxyls

such as amino resins (Fig. 5) (197). In this way peptides are selectively attached at the C-terminus.

This simple procedure provides one of the most useful and reliable means for attaching peptides to resins. Yields are typically greater than 70%, and no prior blocking of the peptide amino groups is required since the large excess of resin amino groups reacts preferentially with the lactone. Repetitive yields on sequencing are around 5% lower than when other coupling methods are used, presumably because the γ-hydroxyl groups can catalyze partial hydrolysis of the homoserine peptide bond in acid.

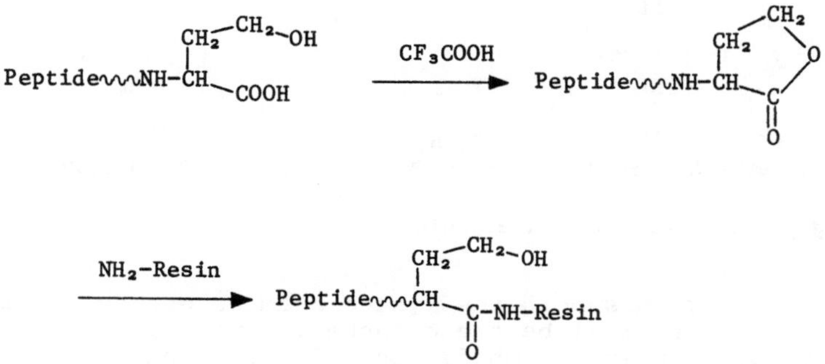

Fig. 5. Homoserine lactone coupling of peptides

In practice the peptide is treated with anhydrous trifluoroacetic acid for 1 h at room temperature, the acid is evaporated, the peptide is neutralized by the addition of triethylamine, and the amino-resin is added. The coupling reaction is carried out in dimethylformamide or a dimethylformamide-water mixture, depending on the solubility of the peptide. If other, nonhomoserine peptides are present, they can be recovered from the resin washings by evaporation of the solvents.

B. Attachment of Amino Side Chains Using *p*-Phenylene Di<u>iso</u>thiocyanate

For peptides which contain lysine or aminoethyl cysteine, the carboxyl activation problem can be circumvented by coupling through the side-chain amino groups (248). The peptide amino groups are activated with *p*-phenylene di<u>iso</u>thiocyanate and coupled with an amino-resin (Fig. 6).

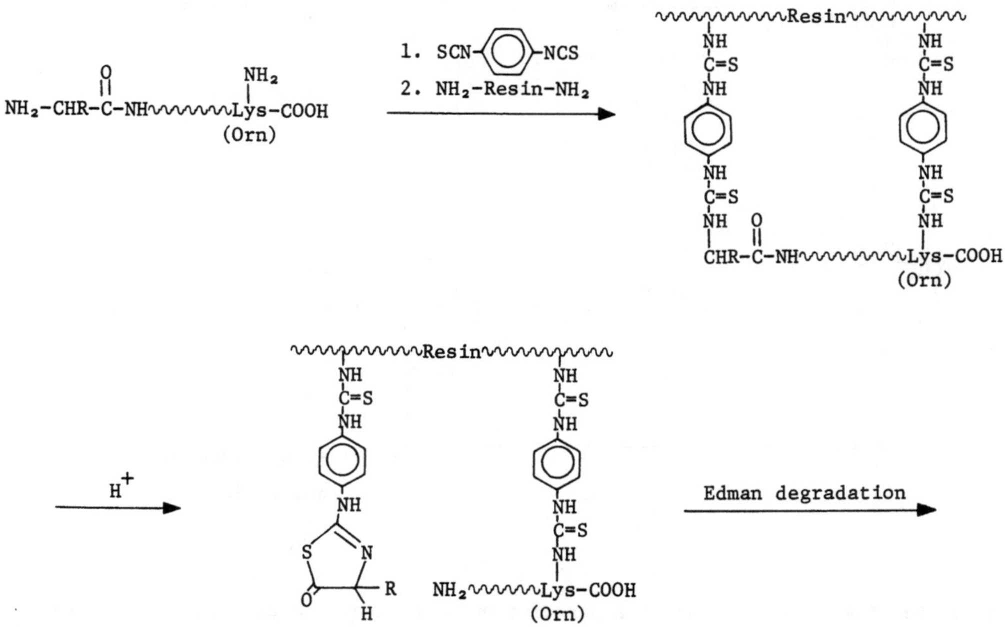

Fig. 6. Di<u>iso</u>thiocyanate coupling of peptides

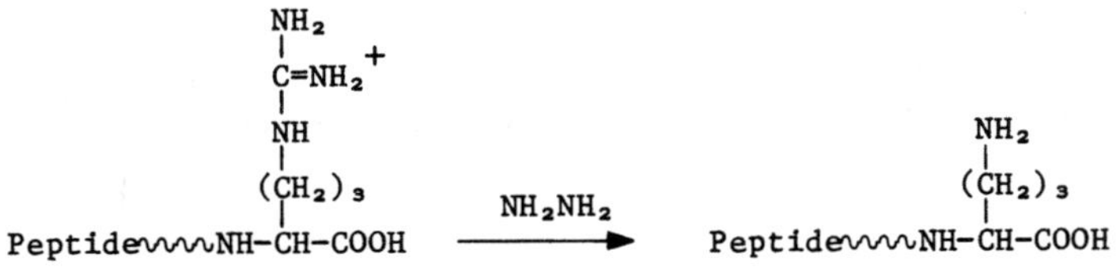

Fig. 7. Deguanidation of arginine

In this process both the lysine ε-amino groups and the α-amino group of the N-terminal become attached. However, treatment of the resin with acid under the conditions of the Edman degradation results in cleavage of the first peptide bond and liberation of a new amino group. Because the phenylthiohydantoin (or thiazolinone) of the first amino acid remains bound to the resin, it must be detected by other means, such as dansylation, or by difference in the composition of the peptide.

This procedure can be carried out in dimethylformamide or pyridine solvent and in the presence of appreciable amounts of salts (e.g., tertiary amine salts) provided a pH of about 9.5 is maintained. It is important to remove primary and secondary amines which can react with the coupling agent. The yields are generally greater than 80%, and the entire procedure can be accomplished in 3-4 h.

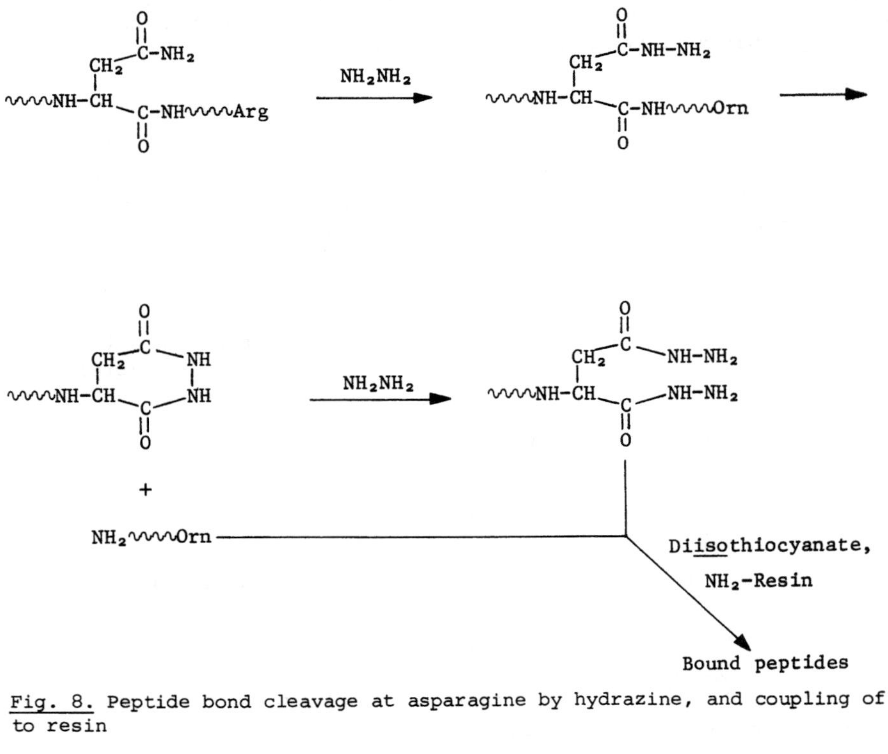

Fig. 8. Peptide bond cleavage at asparagine by hydrazine, and coupling of fragments to resin

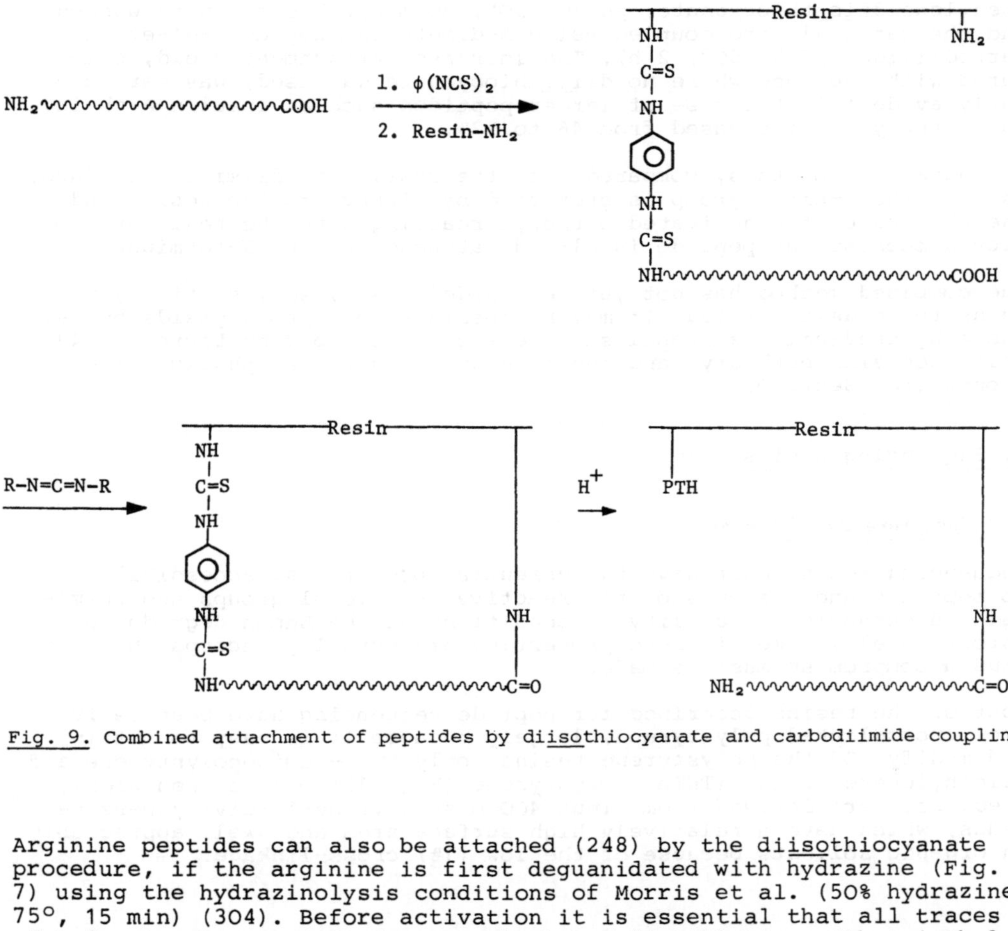

Fig. 9. Combined attachment of peptides by diisothiocyanate and carbodiimide coupling

Arginine peptides can also be attached (248) by the diisothiocyanate procedure, if the arginine is first deguanidated with hydrazine (Fig. 7) using the hydrazinolysis conditions of Morris et al. (50% hydrazine, 75°, 15 min) (304). Before activation it is essential that all traces of hydrazine be removed, for example by co-distillation with triethylamine, since the hydrazine will react with the diisothiocyanate.

Hydrazinolysis leads to several side reactions – cleavage of Arg-Gly peptide bonds (407) and hydrazinolysis of the side chain amide groups of glutamine and asparagine. In the latter case, peptide bond cleavage frequently occurs (Fig. 8) (197, 470) and the fragments become attached to the resin. Schiltz (unpublished) has observed that the peptide Leu-Asp-Asn-Val-Val-Tyr-Arg is cleaved by hydrazine at Asn and that both of the resulting fragments become attached to the resin and are sequenced simultaneously. The number of side reactions is sufficient to make this procedure useful only for smaller arginine peptides or those that do not contain asparagine and glutamine.

C. Combined Methods of Attachment

Although good methods for attachment of peptides with C-terminal lysine or homoserine are available, one is faced with the problem of what to do with peptides that do not contain these amino acids. Schiltz (394) has attacked the problem by using a combination of the diisothiocyanate and carbodiimide methods (Fig. 9). The peptide is first linked to aminopolystyrene (see Sect. D) by its α-amino group in the usual way (*p*-

phenylene diisothiocyanate, pH 10, 50°, 30 min). The resin is washed and the carboxyls are coupled using N-dimethylaminopropyl-N'-ethyl carbodiimide (pH 5, 50°, 2 h). The increased attachment yield, compared with the case where no diisothiocyanate was used, was particularly evident in the case of larger peptides such as insulin A-chain, where the yield increased from 46 to 92%.

The advantage of this, compared with the usual carbodiimide procedure, is that the α-amino group is protected by linkage to the resin, and the chances of the activated carboxyl reacting with the resin are improved because the peptide is already attached at the N-terminus.

The combined method has not yet been widely used, so its utility remains to be demonstrated. It may be possible to improve yields by using the ethylenediamine aminopolystyrene resin described by Laursen (249) which contains both aryl and the more nucleophilic aliphatic amino groups (see Sect. D).

D. Sequencing Resins

1. Polystyrene-Based Resins

Sequencing resins must have two essential properties: accessibility to peptides and reagents of the reactive functional groups and chemical and mechanical stability to conditions of the Edman degradation. Unfortunately, some of these properties are mutually incompatible and thus a compromise must be made.

Most of the resins described for peptide sequencing have been derived from cross-linked polystyrene, largely because it is easy to obtain and modify. Of the polystyrene resins, only two – aminopolystyrene and triethylenetetramine (TETA) polystyrene (Fig. 10) – have been widely used. Both are derived from minus 400 mesh styrene-1%-divinylbenzene beads, which have a relatively high surface area and swell appreciably in organic solvents because of the low (1%) cross-linkage.

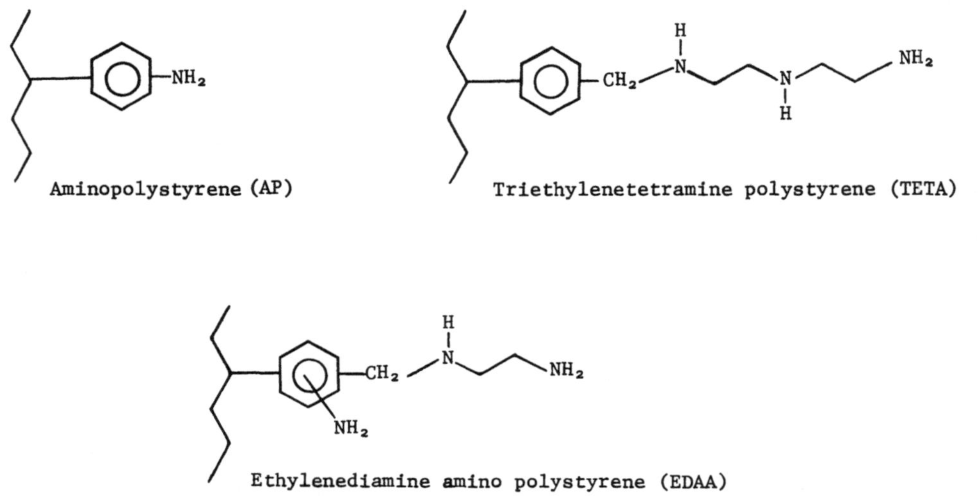

Fig. 10. Peptide sequencing resins

Aminopolystyrene is most useful for attachment of peptides by the diisothiocyanate method, but is less satisfactory for coupling carboxyl-activated peptides. There are two probable reasons for this discrimination. Aminopolystyrene is a hydrophobic resin in its neutral form and is not readily penetrated by peptides, which are usually polar. However, when a peptide is activated with *p*-phenylene di<u>iso</u>thiocyanate, the peptide amino groups acquire reactive hydrophobic "heads" which can readily penetrate and react with the hydrophobic amino resin. Carboxyl-activated peptides, however, are generally polar and cannot penetrate as easily. Also arylamines are not particularly good nucleophiles in carbonyl reactions, and react rather slowly.

On the other hand, triethylenetetramine (TETA) resin, with its long polyamine side chains, is hydrophilic, as seen by its swelling properties in polar solvents such as methanol. The aliphatic amino groups (pKa ca. 6-10) are also better nucleophiles than aryl amino groups. For these reasons the TETA resin gives superior results in coupling carboxyl-activated peptides.

For the combined coupling method, a support such as the EDAA-resin would probably give best results since it has both aryl and aliphatic amino groups. This, however, remains to be tested.

After the peptide has attached to the support, excess resin amino groups are blocked, usually by reaction with methyl <u>iso</u>thiocyanate. The resulting resin methylthiourea is somewhat unstable to acid, with the result that some decomposition products are noted during phenylthiohydantoin (PTH) analysis. Aminopolystyrene gives fairly clean results, while the TETA resin gives rise to a spot eluting with PTH-Thr on thin layer chromotography. Contaminant spots are also found near the origin, but generally do not interfere with analysis. If the TETA resin amino groups are blocked with phenyl<u>iso</u>thiocyanate, the spot near PTH-Thr disappears, but others appear near PTH-Phe and PTH-Val.

It has been observed (unpublished) that attachment yields to TETA decrease after the resin has been stored for several months. This is apparently due to reaction of the activated peptide with unbound amino resulting from decomposition or from the slow release of triethylenetetramine trapped during synthesis. Normal coupling yields can be restored by washing the resin with dimethyl formamide just before use.

Polystyrene-based resins suffer from the problem that they swell and shrink in different solvents, which can lead to channeling or blocking when the resin beads are packed into reaction columns. This difficulty can be avoided by mixing the resin with a large (20-fold) excess of glass beads, so that swelling and shrinking occurs in the interstices between the glass beads (249). Alternatively, reactions can be carried out in a reaction vessel, rather than a column, though this leads to some mechanical problems (140). Preliminary results with rigid pellicular and macroporous resins have not been encouraging so far (70).

2. Porous Glass Supports

Wachter et al. (453) have described the use of porous glass supports for sequencing peptides and proteins (see reference (454) for a review of recent results). These materials are prepared by silylating porous glass beads (Corning CPG 10-75 Å, 200-400 mesh, Electronucleonics, Inc.) with aminopropyl or β-aminoethylaminopropyl siloxanes to give amino-glass supports, which can be used directly or further activated with *p*-phenylene di<u>iso</u>thiocyanate (Fig. 11). Porous glass supports have the

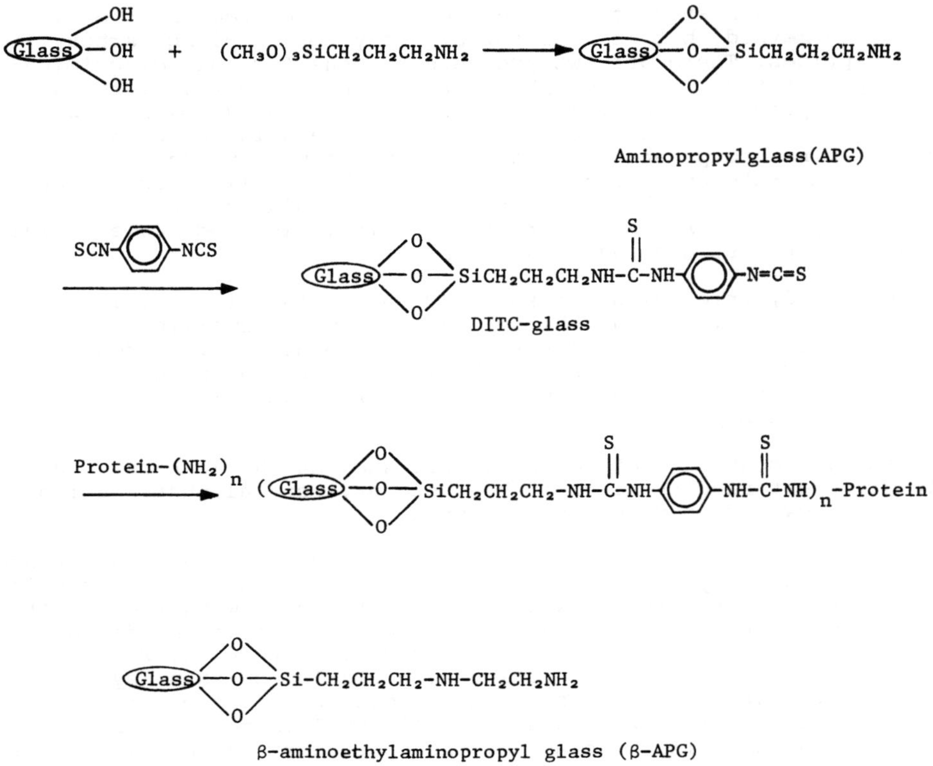

Fig. 11. Amino-glass supports for immobilization of proteins

virtue of being rigid, so that many of the mechanical problems associated with polystyrene resins are avoided. They also have an enormous surface area, so that the reactive groups, which are all on the surfac are very accessible.

Wachter et al. (454) have recently summarized their experience with porous glass supports. For attachment of homoserine peptides, APG (Fig 11) is best for peptides containing 10-40 residues, while β-APG is better for larger peptides. The β-APG appears to be more reactive than APG because of the neighboring amine group which can participate in general base catalysis (Fig. 12) (69, 335). The same effect is seen with TETA polystyrene (197). DITC-glass couples large peptides and proteins in high yields, but gives poor results for small peptides. Apparently the polarity of the peptide is important in this case, since hydrophilic peptides couple less readily than hydrophobic peptides to

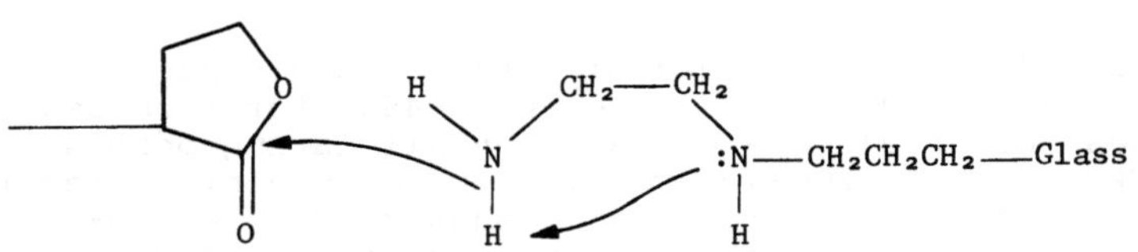

Fig. 12. General base catalysis by vicinal amine

Table 1. Typical combinations of attachment methods and resins

Attachment method	Resin	Peptide type and size
Diisothiocyanate	AP[a]	C-terminal Lys, AECys, Orn; < ∿ 30 residues
	APG, β-APG	C-terminal Lys, AECys, Orn; > 10 residues
	DITC-glass	proteins, large peptides
Homoserine Lactone	TETA	C-terminal Hse; < ∿ 50 residues
	APG	C-terminal Hse; 10-40 residues
	β-APG	C-terminal Hse; > 40 residues
Carbodiimide (EDC or DCC)	TETA, APG, β-APG	All peptides; < ∿ 30 residues
Carbonyldiimidazole	TETA	< ∿ 30 residues

[a]AP: aminopolystyrene; APG: aminopropyl glass; β-APG: β-aminoethylaminopropyl glass; TETA: triethylenetetramine polystyrene; DITC: p-phenylene di<u>iso</u>thiocyanate; EDC: N-ethyl-N'-dimethylaminopropyl carbodiimide; DCC: dicyclohexylcarbodiimide.

the hydrophobic DITC-glass. With all coupling methods, small peptides are attached more efficiently to polystyrene-based resins.

E. Summary of Attachment Procedures

Because solid-phase peptide sequencing techniques are still undergoing development, there is not yet a "best" way to attach peptides to insoluble supports; rather there are a variety of ways, as indicated in Table 1. It is, however, possible to make some generalizations: (a) small peptides are more easily attached to polystyrene resins, while large peptides and proteins couple better to porous glass supports (b) carboxyl-activated peptides couple better to supports with aliphatic amine ligands, and di<u>iso</u>thiocyanate-activated peptides to aminopolystyrene, and (c) the homoserine lactone and di<u>iso</u>thiocyanate coupling procedures are the most reliable and useful.

III. Sequencing Strategies

Given a protein to sequence and solid-phase Edman degradation as the primary sequencing tool, one is faced with the problem of devising an appropriate sequencing strategy. Since Edman degradation of the intact protein generally gives only a small (5 to 10%) portion of the total sequence, it is necessary to fragment the protein pieces of manageable size. By judicious use of the attachment procedures described in the previous section, it is frequently possible to obtain extended sequences of rather large peptides. Furthermore, these selective procedures often permit sequencing of individual peptides in mixtures without prior separation of the components.

```
                      5                   10                  15

 1 SER THR ILE GLU GLU ARG VAL LYS LYS ILE ILE GLY GLU GLN LEU

16 GLY VAL LYS GLN GLU GLU VAL THR ASP ASN ALA SER PHE VAL GLU

31 ASP LEU GLY ALA ASP SER LEU ASP THR VAL GLU LEU VAL MET ALA

46 LEU GLU GLU GLU PHE ASP THR GLU ILE PRO ASP GLU GLU ALA GLU

61 LYS ILE THR THR VAL GLN ALA ALA ILE ASP TYR ILE ASN GLY HIS

76 GLN ALA
```

↓ BrCN

Ser_1----------Arg_6----------Lys_8Lys_9----------Lys_{18}----------------Hse_{44}

\+

Ala_{45}-----------------------------------Lys_{61}---------------------Ala_{77}

Fig. 13. Sequence and BrCN peptides of acyl carrier protein

Although solid-phase Edman degradation has been used to obtain continuous sequences for as many as 50 residues (273), 25-30 cycles appears to be the practical limit in peptide sequencing. This being the case, it would seem desirable to generate small peptides containing fewer than 30 amino acids. In general, however, fragmentation with trypsin or chymotrypsin, methods which produce small peptides, is impractical, since the isolation, purification and alignment of large numbers of small peptides is often time-consuming, tedious, and difficult. Alternately, larger peptides may be generated by cleavage of the protein with cyanogen bromide. Other large peptides may be generated by tryptic cleavage of the protein at arginine, following citraconylation (25) or maleylation (75) of the lysines, or at lysine after blocking of the arginines with cyclohexanedione (337).

Since each protein or peptide presents different sequencing problems, no single strategy can be proposed which is universally applicable. Therefore, several hypothetical examples will be presented to demonstrate the versatility of solid-phase sequencing in the sequence determination of large peptides and proteins.

Acyl carrier protein from *E. coli* E-26 contains 77 amino acid residues including 1 methionine, 1 arginine and 4 lysines (389). This protein, despite its length, can in principle be sequenced by solid-phase Edman degradation without the necessity of any chromatographic separation of peptide fragments (Figs. 13-15). After digestion of the protein with cyanogen bromide, the peptide mixture can by subjected to coupling via the homoserine lactone procedure. Only the N-terminal fragment containing homoserine will become attached to the resin; sequencing will give the first 25-30 residues of the protein (Fig. 14). In another experi-

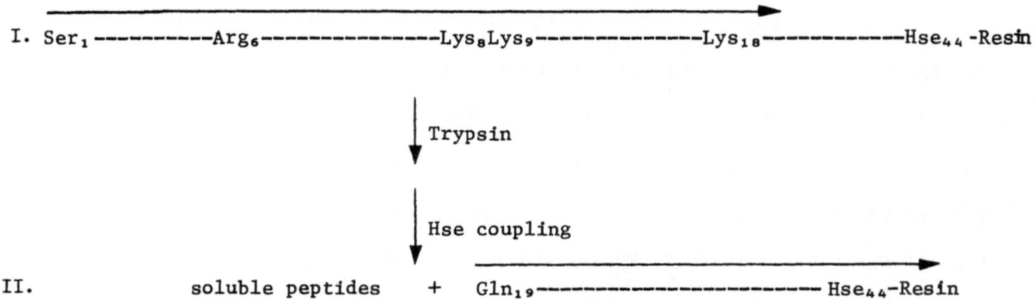

Fig. 14. Hypothetical strategy for sequencing the N-terminal BrCN fragment of acyl carrier protein. I the peptide is coupled and the N-terminal segment sequenced. II a sample of peptide is cleaved with trypsin and the C-terminal segment is coupled and sequenced

ment, the cyanogen bromide-digested protein can be further digested with trypsin, and the resulting peptide mixture coupled again using the homoserine lactone method and sequenced. In this way, residues 19-44 can be determined and this information combined with previous data to give the sequence of the first 44 residues.

It has been shown (198) that the homoserine coupling method can be used to obtain nearly quantitative removal of C-terminal homoserine peptides from peptide mixtures. Thus, attachment of the homoserine-containing peptide from the original CNBr digest of the protein would be expected to leave the single C-terminal peptide (residues 45-77 of the protein) unattached and relatively pure. After attachment by the di<u>iso</u>thiocyanate method, this peptide can be sequenced through its lysine (residue 61) at which point the remaining peptide will become attached from the resin. The detached peptide can be recovered, coupled to a resin by the carboxyl groups and sequenced to give the remaining 16 residues, 62-77

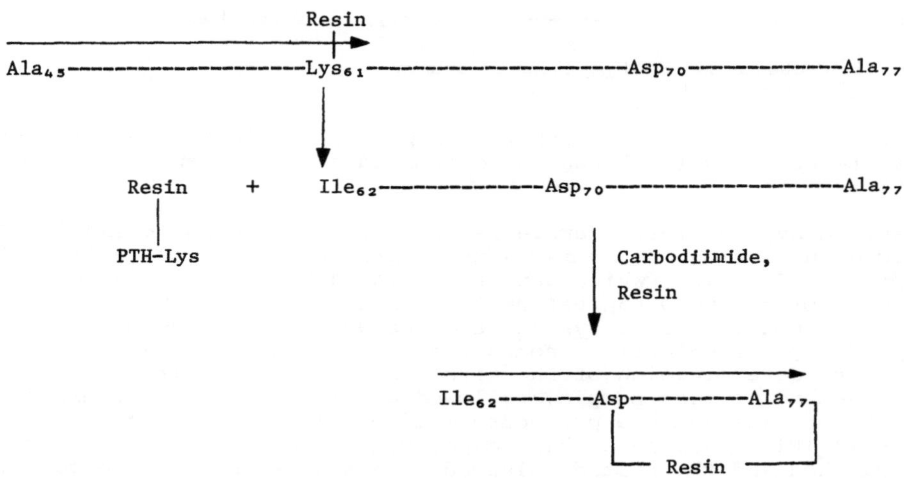

Fig. 15. Hypothetical strategy for sequencing the C-terminal BrCN fragment of acyl carrier protein. The peptide is attached at lysine and sequenced through Lys_{61}. The detached C-terminal segment is reattached and sequenced

					5					10					15
1	ALA	GLU	CYS	SER	VAL	ASP	ILE	ALA	GLY	THR	ASP	GLN	MET	GLN	PHE
16	ASP	LYS	LYS	ALA	ILE	GLU	VAL	SER	LYS	SER	CYS	LYS	GLN	PHE	THR
31	VAL	ASN	LEU	LYS	HIS	THR	GLY	LYS	LEU	PRO	ARG	ASN	VAL	MET	GLY
46	HIS	ASN	TRP	VAL	LEU	THR	LYS	THR	ALA	ASP	MET	GLN	ALA	VAL	GLU
61	LYS	ASP	GLY	ILE	ALA	ALA	GLY	LEU	ASP	ASN	GLN	TYR	LEU	LYS	ALA
76	GLY	ASP	THR	ARG	VAL	LEU	ALA	HIS	THR	LYS	VAL	LEU	GLY	GLY	GLY
91	GLU	SER	ASP	SER	VAL	THR	PHE	ASP	VAL	ALA	LYS	LEU	ALA	ALA	GLY
106	ASP	ASP	TYR	THR	PHE	PHE	CYS	SER	PHE	PRO	GLY	HIS	GLY	ALA	LEU
121	MET	LYS	GLY	THR	LEU	LYS	LEU	VAL	ASP						

↓ BrCN

Ala$_2$------------------------------------Hse$_{13}$ 13 residues

Gln$_{14}$-----------------------------------Hse$_{44}$ 30 residues

Gly$_{45}$-----------------------------------Hse$_{56}$ 12 residues

Gln$_{57}$-----------------------------------Hse$_{121}$ 65 residues

Lys$_{122}$----------------------------------Asp$_{129}$ 10 residues

Fig. 16. Sequence and BrCN peptides of azurin

(Fig. 15). Residue 45, which will appear as a blank after diisothiocyanate coupling (248) can be determined by dansyl and group analysis (165).

A second hypothetical example is outlined in Figures 16 and 17. Cyanoge bromide digestion of *Bordetella bronchiseptica* azurin (12) will yield five peptides (Fig. 16), which can be separated by conventional means. The largest of these, a 65 residue C-terminal homoserine peptide (residues 57-121), contains four lysines and one arginine. Attachment of this peptide by its C-terminal homoserine or the internal lysines will allow sequence determination of the first 25-30 residues. Tryptic digestion of a second peptide sample, followed by attachment of the mixture via the homoserine lactone procedure will permit the sequencing of residues 102-121 (Fig. 17). In a third experiment, the peptide lysines can then be blocked and the peptide cleaved at arginine. Homoserine coupling of the peptide mixture and sequencing of the single attached peptide shoul provide sufficient information to overlap the N- and C-terminal portion of the peptide. The remaining cyanogen bromide peptides, which range

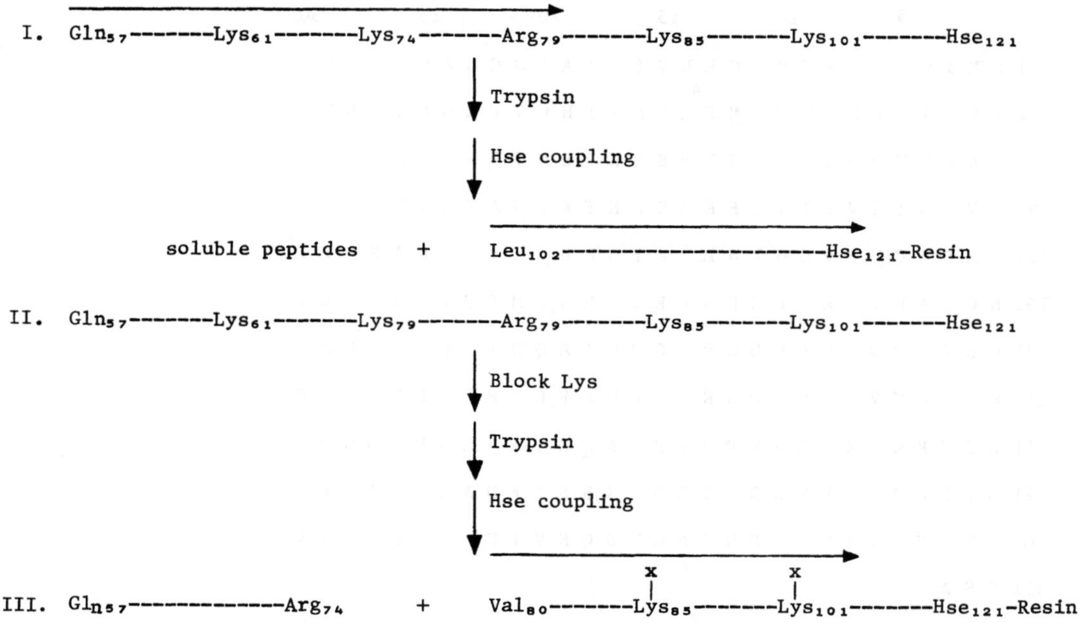

Fig. 17. Sequencing strategy for azurin peptides 57-121. I the intact peptide is coupled and partially sequenced through about 20 residues. II a sample of the peptide is cleaved with trypsin, and the C-terminal segment is selectively attached and sequenced. III a sample of the peptides is cleaved at arginine and is selectively attached and sequenced

in length from 10 to 30 residues, can be sequenced either in a single run or by using the strategy for residues 57-121.

The examples above also illustrate one of the pitfalls encountered in sequencing. Some of the largest peptides have N-terminal glutamine, which can be expected to cyclize to pyroglutamic acid, thus making Edman degradation impossible. Nevertheless some C-terminal sequence data can be obtained using the strategy outlined. Alternatively one may be able to deblock the N-terminal using pyroglutaminase (123).

A third example is provided by lobster glyceraldehyde-3-phosphate dehydrogenase, a 333-residue protein containing 10 methionine residues (Fig. 18). Since the N-terminal of the protein is acetylated (105), sequencing of the intact protein is precluded. The five smaller peptides (3-23 residues) can be sequenced directly, the two containing 31 residues and that with 56 residues using the strategies discussed above. These procedures should yield about 60% of the sequences of the N-blocked peptide, and of the 50- and 69-residue peptides. In all, one might expect to obtain about 85% of the cyanogen bromide peptide sequences without purification of subfragments.

Selective coupling of peptides can also be used to separate arginine and lysine peptides. As illustrated in Figure 19 (p. 31), a large peptide isolated after reversible blocking of arginine and tryptic cleavage can be partially sequenced. A second sample can be cleaved at arginine by trypsin, and the mixture of peptides attached by the diisothiocyanate method. Both peptides will become attached – one by the

```
          5         10        15        20        25        30

  1 S K I G I D G F G R I G R L V L R A A L S C G A Q V V A V N
 31 D P F I A L E Y M V Y M F K Y D S T H G V F K G E V K M E D
 61 G A L V V D G K K I T V F N E M K P E N I P W S K A G A E Y
 91 I V E S T G V F T T I E K A S A H F K G G A K K V V I S A P
121 S A D A P M F V C G V N L E K Y S K D M T V V S N A S C T T
151 N C L A P V A K V L H E N F E I V E G L M T T V H A V T A T
181 Q K T V D G P S A K D W R G G R G A A Q N I I P S S T G A A
211 K A V G K V I P E L D G K L T G M A F R V P T P D V S V V D
241 L T V R L G K E C S Y D D I K A A M K T A S E G P L Q G F L
271 G Y T E D D V V S S D F I G D N R S S I F D A K A G I Q L S
301 K T F V K V V S W Y D N E F G Y S Q R V I D L L K H M Q K V
331 D S A
```

↓ BrCN

	Number of Residures
Ac-Ser$_{1}$------------------Hse$_{39}$	39
Val$_{40}$-----------------Hse$_{42}$	3
Phe$_{43}$-----------------Hse$_{53}$	11
Glu$_{54}$-----------------Hse$_{76}$	23
Lys$_{77}$-----------------Hse$_{126}$	50
Phe$_{127}$----------------Hse$_{140}$	14
Thr$_{141}$----------------Hse$_{171}$	31
Thr$_{172}$----------------Hse$_{227}$	56
Ala$_{228}$----------------Hse$_{258}$	31
Lys$_{259}$----------------Hse$_{327}$	69
Gln$_{328}$----------------Ala$_{333}$	6

Fig. 18. Sequence and cyanogen bromide peptides of lobster glyceraldehyde-3 phosphat dehydrogenase

N^{α}- and N^{ε}-amino groups, the other by N^{α}- only. On treatment of the resin with acid, the arginine peptide will be released (minus its N-terminal) and the lysine peptide alone will be sequenced. Variations of this strategy include sequencing of lysine peptides contaminated

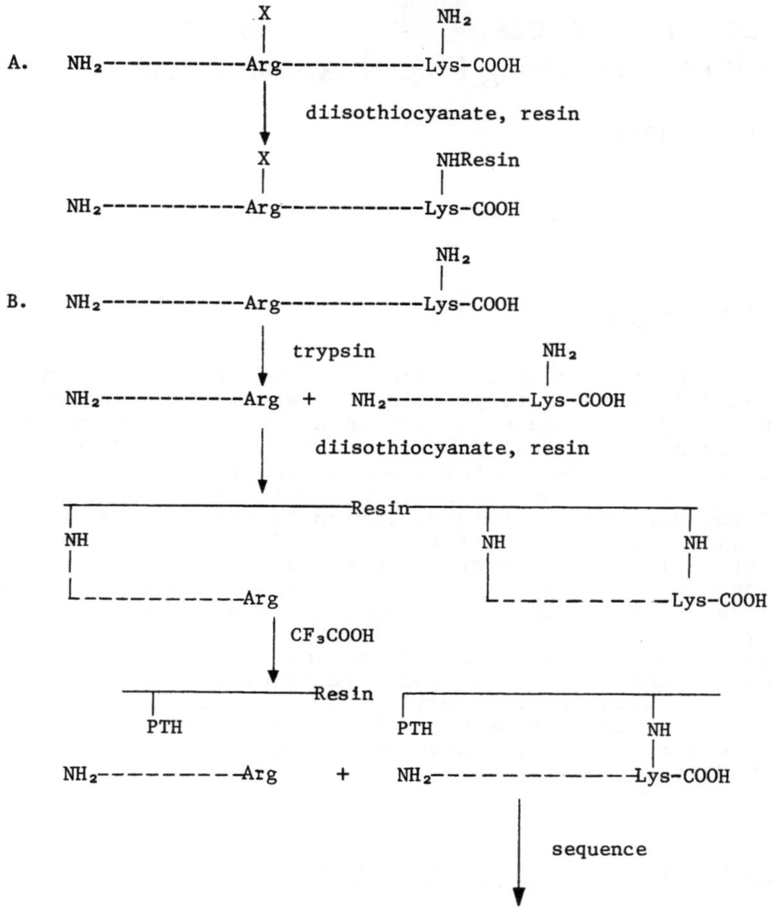

Fig. 19. Sequencing strategy for peptides containing lysine and arginine. A a peptide with C-terminal lysine is attached and partially sequenced. B the peptide is cleaved with trypsin at arginine and the lysine segment is attached and selectively sequenced; the arginine peptide falls off the resin in the first step of the Edman degradation

with nonlysine peptides, and subtractive Edman degradation of non-lysine peptides.

Other sequencing strategies are possible, depending on the amino acid composition of the peptide and the ingenuity of the sequencer. The utility of the solid-phase technique will, one hopes, be further increased by the development of new selective methods for peptide attachment and of an improved general procedure for coupling carboxyl groups to insoluble supports.

Acknowledgment. Some of the work described here was supported by a grant to R.A.L. from the National Science Foundation.

Sequencing Peptides and Proteins Lacking Free α-Amino Groups

R. F. DOOLITTLE

I. Introduction

Facing a protein with a blocked α-amino group, even the Automatic Sequenator (133, 134) is powerless, thwarted by a Nature unappreciative of the universal importance of amino-terminal step-wise degradation procedures. In general, these recalcitrant materials fall into two groups: those in which the α-amino groups are acylated – usually with an acetyl group – and those in which the terminal amino acid is a pyrrolidone carboxylyl residue, formed by the cyclization of a terminal glutaminyl or glutamyl residue. In addition, there are a few instances of naturally occurring cyclic peptides which also lack free α-amino groups. The frustrations which accrue when working with these substance include difficulties in finding them, since they do not react with ninhydrin or fluorescamine (unless the peptide contains one or more lysyl ε-amino groups), as well as identifying the terminal residues, since all the traditional end group procedures employing FDNB, DNS-Cl, NCO, etc., also are ineffective. In this article some simple methods for finding peptides with blocked α-amino groups are reviewed, as well as some strategies presented for identifying and/or removing terminal blocked residues.

II. Isolation and Detection of Peptides Lacking Free α-Amino Groups

How difficult it may be to find a peptide with a blocked α-amino group will depend on a number of factors, the most critical of which is usually the amounts of material available for investigation. In this chapter we are presuming that the investigator is operating in the 1-100 nmol range. Given that to be the case, the location and/or isolation of ninhydrin-negative peptides can be accomplished either with paper or thin layer supports, or with column chromatographic procedures.

Useful Stains on Paper or Thin Layers. Ninhydrin- or fluorescamine-negative peptides can be detected on paper or thin layer supports by a variety of other stains, depending on the amino acid composition of the peptide under study. Among the most useful of these is the arginine stain of Yamada and Itano (476) which is exceedingly sensitive and the method of choice if arginine is present. The classic Pauli stain for histidine (415) also is very good, and tryptophan-containing peptides are often fluorescent (without staining) or can be detected with the Ehrlich stain (415). Other ninhydrin-negative peptides can be detected with the chlorine gas method (369), the sensitivity of which depends roughly on the number of residues in the peptide, the more the better. Finally, there is the brute force approach whereby every sector of a paper strip is eluted, hydrolyzed and analyzed on an amino acid analyzer. This method is tedious but effective.

Column Chromatography Procedures. Investigators employing autoanalyzers with ninhydrin or fluorescamine detectors frequently miss peptides with blocked α-amino groups if the peptides are not hydrolyzed as a part of the procedure. Beyond that, the problem is usually more one of finding the single blocked peptide representing the amino terminus among all the others in an enzymic digest of a given protein. To this end, Narita (315) devised an ion exchange approach which takes advantage of the reduced positive charge associated with the absence of an α-amino group. Thus, if an appropriate enzymatic digest of a blocked protein is applied to Dowex 50 at low pH, only the most electronegative peptides flow through unadsorbed. The key here is to know in advance what is "an appropriate enzymatic digest". Pepsin, chymotrypsin, subtilisin and pronase have been the most popular choices (196, 209, 315), since trypsin digestion results in virtually every peptide having a carboxy-terminal arginine or lysine, the positive charges of which will bind to the Dowex 50. Trypsin can be used effectively, however, if the digest is subsequently incubated with carboxypeptidase B to remove the carboxy-terminal arginines and lysines. In any case, care should be taken to acidify the digest before application to the Dowex 50 column. Minimization of artifactual pyrrolidone carboxylic acid formation can be achieved by performing all operations in the cold (209).

III. PCA-Terminating Peptides and Proteins

Terminal pyrrolidone carboxylic acid (PCA) residues (I) are the result of the cyclization – either spontaneous or enzymatically catalyzed – of glutaminyl (II) or glutamyl (III) amino-terminals (51).

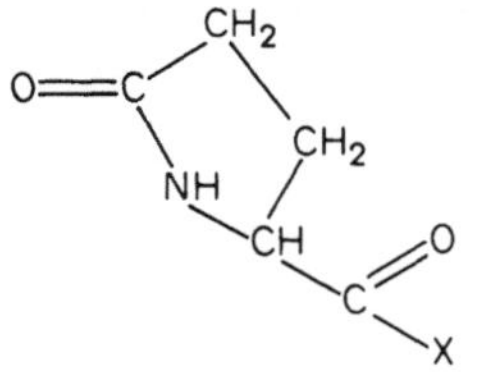

(I)

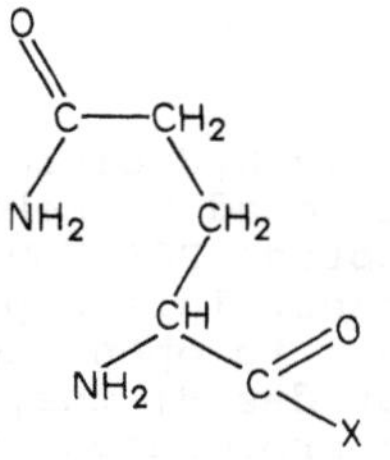

(II)

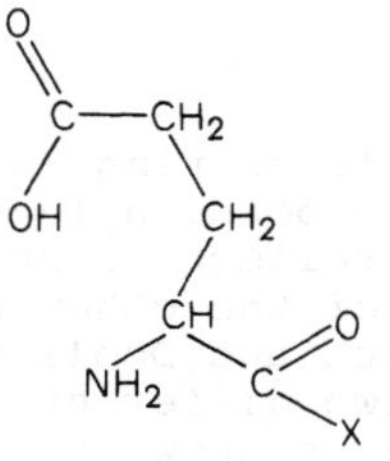

(III)

The groups, which are sometimes referred to as pyroglutamyl residues, occur naturally in a variety of peptides and proteins (Table 1), as well artifactually as by-products brought about during the fragmentation and purification of peptides. In general, there are two approaches to their determination: (a) chemical modification, and (b) selective enzymatic removal of the residue.

Table 1. Some naturally occurring proteins and peptides known to have terminal pyrrolidone carboxylic acid[a]

Fibrinogen β-chains (assorted species)
Immunoglobulin γ-chains (assorted species)
Immunoglobulin μ-chains (assorted species)
Immunoglobulin α-chains (assorted species)
Immunoglobulin λ-chains (assorted species)
Orosomucoid (α_1 acid glycoprotein) (human)
Cold insoluble globulin (human)[b]
High density lipoprotein (human)
Collagen (assorted species)
Cytochrome b_5 (rabbit)
Prophospholipase (pig)
Gastrin (assorted mammals)
Releasing factors (assorted mammals)
Snake venom peptides (assorted peptides in assorted species)
Amphibian secretions (assorted)
Insect venoms (assorted)
Blanching hormone (prawn)
Molluscan venoms (octopus)
Fastigiatin[c] (algae)

[a]Taken from Dayhoff's Atlas (108, 109).

[b]Personal communication from M. Mosesson.

[c]Fastigiatin is a tripeptide, PCA-GLN-GLN (112).

A. Chemical Approaches

1. Alkali Lability

The amide linkage in the pyrrolidone ring is more readily hydrolyzed by base than the average peptide bond, a feature first noted by Dekker et al. (112) who opened the pyrrolidone ring the tripeptide PCA-GLN-GLN without significant hydrolysis of the other peptide bonds. This approac was subsequently used by Blombäck and Doolittle (52, 53) to open the terminal pyrrolidone rings of two different fibrinopeptides B; they were then able to apply ordinary step-wise degradation procedures to obtain the amino acid sequences. Under the conditions used (1N NaOH at room temperature for 24 h) the yields obtained were only about 30%, however. If more vigorous conditions were applied, indiscriminate peptide bond cleavage rendered the method virtually useless. They also reported that N-acetylated amino acids were stable under the same alkaline conditions which open the pyrrolidone ring (52), and the alkali lability test was subsequently used as an aid in identifying pyrrolidon carboxylic acid at the amino-terminal immunoglobulin γ-chains (352).

2. Reductive Conversion

Takahashi and Cohen (425) reported that terminal pyrrolidone carboxyl residues could be reduced with diborane (in tetrahydrofuran or tetramethylurea) and effectively converted into proline residues. The maximum yield reported for several simple peptides investigated was 46% (269). The method has a number of disadvantages, including the fact that carboxyl groups are reduced significantly faster than the pyrrolidone ring amide; moreover, a substantial amount of reduction occurs among ordinary peptide bonds, thereby precluding further sequence studies by traditional procedures.

3. Methanolysis

The fact that many peptides and proteins can be successfully esterified with methanolic-HCl without rupture of peptide bonds led Kawasaki and Itano (229) to undertake an investigation of the effects of this treatment on the pyrrolidone ring of a variety of small PCA-peptides. They found that 1N HCl/methanol at 25-40° for 1-3 days was quite effective in converting the pyrrolidone residues to the γ-methyl ester of glutamic acid. Although carboxyl groups are speedily esterified under these conditions, most peptide bonds are unaffected, and step-wise degradation procedures on the products ought to be possible. Peptide bonds involving glycine are also unusually sensitive to methanolysis, however, thus somewhat limiting the usefulness of the approach (229). On the other hand, the method is simple and may be put to good advantage on small peptides lacking glycine.

4. Mass Spectrometry

The determination of terminal pyrrolidone carboxylic acid residues by mass spectrometry is a rapid and simple procedure for investigators with access to appropriate instruments. Lenard and Gallop (252) used a Hitachi-Perkin Elmer RMU-6E mass spectrometer to identify the PCA in a synthetic hexapeptide which had first been N-methylated (silver oxide and methyl iodide in dimethyl formamide at room temperature) to increase its volatility. A unique fragment ion with m/e = 98 was obtained in good yield as a result of the cleavage of the C_{α}-C=O bond. DeHaas et al. (111) were able to identify PCA positively as the terminal residue in a heptapeptide from porcine phospholipase A after derivatization with CD_3I; in this case the corresponding fragment had an m/e = 101 (Fig. 1).

Sometimes PCA-peptides are sufficiently nonpolar that prior modification is unnecessary. Thus, Chang et al. (87) studied the synthetic tripeptide PCA-HIS-PRO-NH_2 (pyrrolidone carboxylyl-histidyl prolinamide) directly and obtained a characteristic fragment of m/e = 84 with both a low resolution instrument (CEC 21-491) and a high resolution device (CEC 21-110). The same particle was found by Fernlund and Josefsson (151) in their investigation of the prawn blanching hormone, an octapeptide with no free carboxyl or amino groups. These authors used an AEI-MS 902, and only 6 μg of sample was required. Mass spectrometry is clearly an excellent approach for the positive identification of pyrrolidone carboxylic acid residues, and its use ought to become increasingly popular.

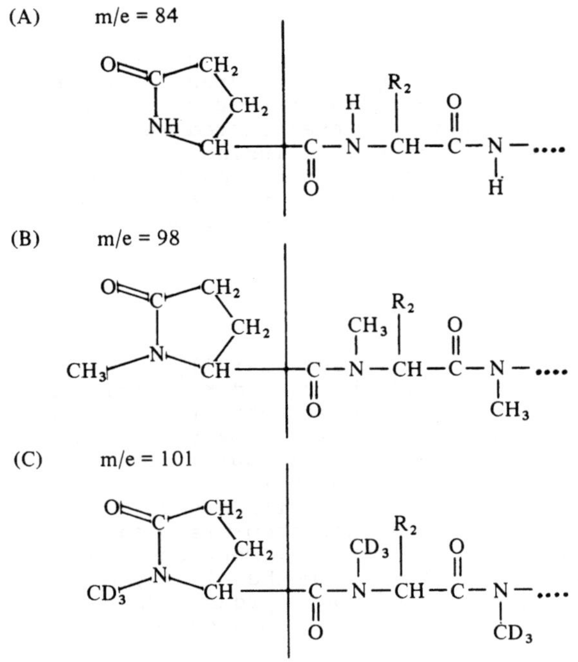

Fig. 1A-C. Characteristic fragment ions observed by mass spectrometry of (A) underivatized, (B) methylated, and (C) deuteromethylated PCA-peptides

5. Guessing

Although hardly a chemical approach, guessing that a blocked amino-terminal residue is a pyrrolidone carboxylic acid is not uncommon. What generally happens is that no end group is found by standard procedures. These negative observations are then followed up by another null set, no acyl groups being found by methods which are frequently not so standard for the peptide chemist (Sect. IV-A). Amino acid analysis after total acid hydrolysis reveals the presence of glutamic acid in the terminal peptide, reinforcing a preconceived notion. If homologous peptides already have been reported as having terminal PCA, there is hardly an investigator at work with restraint enough not to make it a foregone conclusion.[1]

About the only time it is really legitimate to guess, however, is when the substance under study is biologically active and can be synthesize Thus, in both the cases of eledoisin (13) and gastrin (168) it was guessed that the blocked amino-terminals must be PCA. That these were correct guesses was borne out by the total synthesis of the peptides and the demonstration that the PCA-peptides were in fact biologically active (16, 59).

[1]The risk of the guess involved is usually inversely proportional to the number of times words and phrases such as "unequivocally" or "proves without doubt" are employed in describing the negative results.

B. Enzymatic Removal of PCA Residues

Pyrrolidone carboxylyl residues can be preferentially cleaved from the termini of peptides and proteins by the enzyme pyrrolidone carboxylyl peptidase (originally called pyrrolidonyl peptidase). The enzyme was first isolated from a selected strain of *Pseudomonas fluorescens* (120), but it can be prepared from a variety of bacterial and other sources (20, 121, 422, 423). The recipes described in this section all pertain to the enzyme isolated from the original source strain (which can be obtained from the American Type Culture Collection, Rockville, Maryland, U.S.A., Strain No. 25289), but most of the comments also seem to apply to the enzyme isolated from other sources (119).

1. Assay Method

PCA-peptidase is not a particularly stable enzyme, and as such it is important that the investigator trying to cleave a terminal PCA from some valuable peptide or protein be certain that an active enzyme preparation is actually being employed. To this end, even the recipient of a gift of enzyme for a "once only" experiment ought to be prepared to assay the enzyme under defined conditions. The standard assay used in our laboratory depends on the enzyme-catalyzed hydrolysis of L-pyrrolidone carboxylyl L-alanine (PCA-ALA). The progress of the reaction is followed by monitoring the exposure of the alanine α-amino groups with ninhydrin.

A stock solution of PCA-ALA (MW = 200) containing 25 μmol/ml (5.0 mg/ml) is prepared in glass distilled water. Ordinarily, 50 μl of buffered (pH = 7.3) enzyme solution is added to 25 μl of the stock substrate solution, the final PCA-ALA concentration being $8.3 \cdot 10^{-3}$ M. Digestions are carried out at 30° in stoppered tubes. At appropriate times 1.0 ml absolute ethanol is added to the mixture to poison the enzyme and precipitate the bulk of the protein. The stoppered tubes are placed in the cold for at least 15 min to aid precipitation and then centrifuged in a table-top centrifuge. The supernatants are decanted into clean tubes and 0.5 ml aliquots removed for ninhydrin assay. At this point 0.5 ml of ninhydrin reagent is added; the tubes are then capped loosely, and placed in a boiling water bath for exactly 15 min (191). After cooling, the solutions are diluted with 5 ml of 50% isopropanol and the absorbance read at 570 mμ. The ninhydrin reagent is made up according to Hirs et al. (191), 2.0 g of ninhydrin and 0.3 g of hydrindantin being dissolved in 75 ml ethylene glycol monomethyl ether (methyl cellosolve), followed by addition of 25 ml of 4 M sodium acetate buffer, pH 5.5. Under these conditions a leucine standard of 100 nmols corresponds to about 0.40 absorbance units. If all the PCA-ALA in a digestion tube is hydrolyzed, a nearly full scale deflection of 1.2 absorbance units above background is obtained. In crude extracts of the enzyme there is frequently a significant exposure of amino groups which is independent of the presence of added substrate and presumably results from endogenous proteolysis. As a control, then, a separate set of incubation mixtures is set up without substrate (PCA-ALA) but containing instead an equivalent amount of pyrrolidone carboxylic acid (PCA).

A unit of enzyme activity is defined as the amount of enzyme which releases one mμmol of alanine per min under the conditions described above, the pH being maintained at 7.3 with phosphate buffers. We have found it convenient to calculate relative specific activities using the absorbance at 280 mμ (A_{280}) of the enzyme solutions as an index of protein concentration.

Szewczuk and Mulczyk (423) use a colorimetric assay employing L-pyrrolidonyl-β-naphthylamide (available commercially from Cyclo Chem. Corp.) as a substrate and measure the released β-naphthylamine by the Bratton-Marshall reaction (66). We have also tried to effect a more direct assay by using L-pyrrolidonyl nitroanilide and measuring the yellow color of freed nitroaniline at 390 mμ (369). Although the enzyme is active towards this substrate, the poor solubility of the material in water at neutral pH (ca. 1 mg/ml) limits its usefulness. We still find the ninhydrin assay and PCA-ALA to be the most sensitive and convenient. Several PCA-dipeptides, including PCA-ALA, are available commercially (Cyclo Chem. Corp.). In addition, we have published a complet description of the synthesis and properties of several others (443).

2. *Preparation of PCA-Peptides from the Original Strain of Pseudomonas Fluorescens*

The enzyme can be prepared either from bacterial pellets frozen directly after harvesting, or from freeze-dried suspensions. The latter material is especially useful for shipping purposes because of its intrinsic stability (freeze-dried cells still retain high levels of activity after seven years). In any case, cell disruption is readily accomplished by sonication after suspension of the cells in a 0.05 M phosphate buffer, pH 7.3, containing 0.01 M mercaptoethanol and 0.001 M EDTA. This buffer, which is used throughout most of the early purification steps, should also contain 0.1 M 2-pyrrolidone as an enzyme stabilizer (21). Care must be taken to maintain the temperature of the suspension during sonication in the range 0-10°.

After sonication the material is centrifuged at 39,000 *g* at 0° for 30 - 60 min. The supernatant liquid containing the enzyme activity is carefully decanted, and if necessary, it should be centrifuged a second time. The A_{280} of the supernatant should be at least 50; if it is higher, it can be diluted with the phosphate buffer described above.

Nucleic acids may be precipitated by adding one volume of 1% protamine sulfate to six volumes of cold supernatant containing the enzyme. Although this step is not absolutely necessary and was not used in our early preparations, the removal of nucleic acids does seem to facilitat the subsequent gel filtration step of Sephadex G-200. After centrifugation (15,000 *g* for 20 min), the pellet is discarded and the supernatant used immediately for ammonium sulfate fractionation.

A saturated ammonium sulfate solution is added directly, with stirring, to the supernatant fluid from the protamine sulfate step until 42% saturation is achieved. After standing at 0° for 1 h, the preparation is centrifuged (39,000 *g* for 30 min). The supernatant fluid is carefully decanted and discarded, the walls of the tubes being wiped dry with tissues. The pellets are dissolved in a minimal volume of phosphate buffer and diluted to an A_{280} of approximately 250. The material is stored frozen.

The next step in the purification entails gel filtration on Sephadex G-200. A 5 × 50 cm column is suitable for large scale purifications; the enzyme activity comes behind the main protein peak. After removal of aliquots for assay and absorbance measurement, the enzyme activity is precipitated from the pool by the addition of solid ammonium sulfate (0.5 g/ml). After at least 1 h at 0°, the material is centrifuged and the pellets stored frozen. If no further purification is to be undertaken, the slurried pool is divided into several portions before centrifugation and the subpellets frozen separately (Table 2). Enzyme preparations at this stage are generally quite suitable for digestion

Table 2. Recommended isolation scheme for pyrrolidone carboxyl peptidase

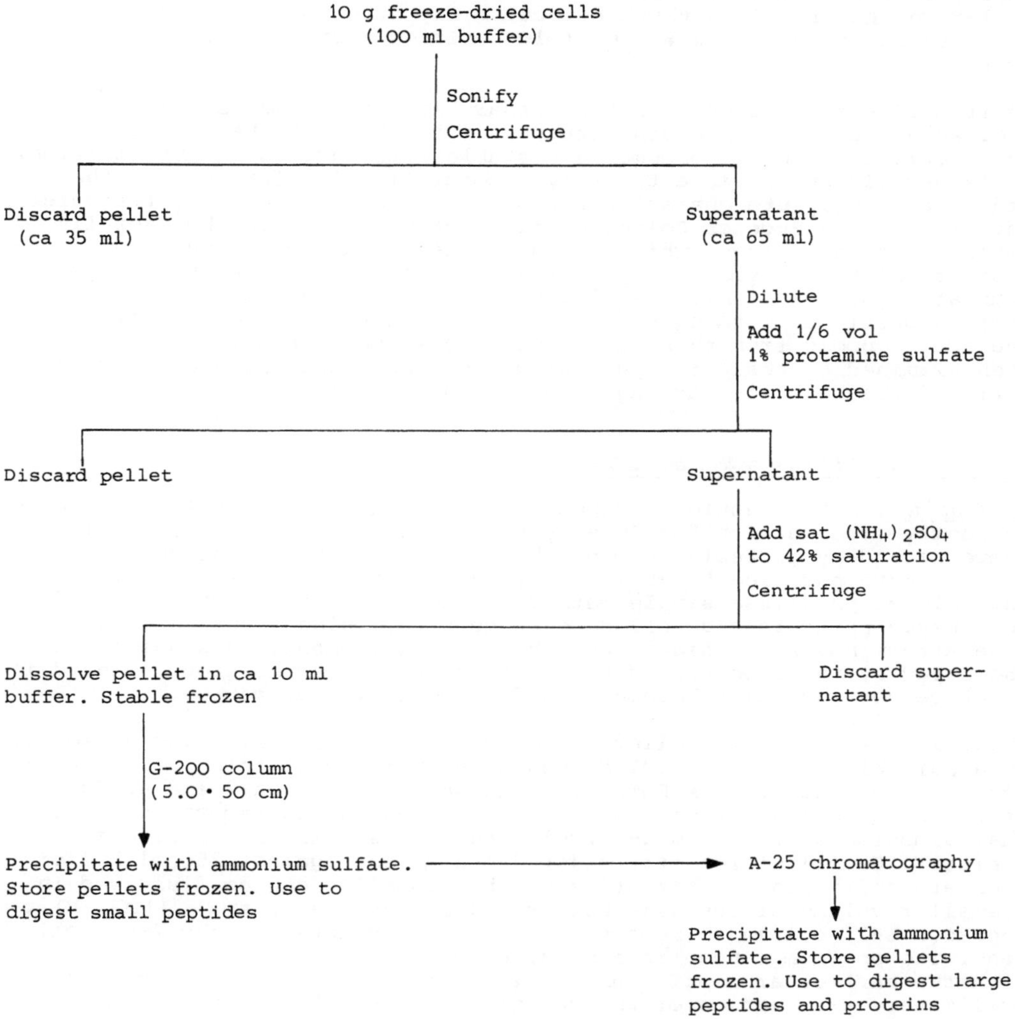

of small PCA-peptides, having a specific activity 20-40-fold that of the original crude extract (sonicate supernatant).

Further purification of the enzyme can be accomplished by ion-exchange chromatography on DEAE-Sephadex (A-25). The pooled pellets from G-200 columns are dissolved in a minimal volume of 0.05 M phosphate buffer, pH 8.0, and desalted either by rapid dialysis or gel filtration on Sephadex G-25. The enzyme activity is readily eluted at pH 8.0 (0.05 M) using a 0-3% NaCl gradient. The active tubes, which come early in the gradient, are pooled and solid ammonium sulfate added (0.6 g/ml) to precipitate the enzyme. Once again, if no further purification steps are planned, it is convenient to divide the slurry into a half-dozen portions before centrifuging. The pellet or subpellets are stored frozen at -70°. Enzyme preparations at this stage of purification have been employed successfully in the removal of PCA residues from a large molecular weight protein without detecting any other proteolysis (21). The material ranges from 60-100-fold purified over the crude bacterial extract.

Additional purification is not necessary if the enzyme is only to be used as a reagent in protein structure work. In fact, because the enzyme activity tends to become less stable with increasing purification, it is usually undesirable to carry the purification further. On the other hand, complete characterization of the enzyme necessitates purer material, and a Buchler Polyprep gel electrophoresis can be usefully employed at this stage. This approach takes advantage of the great mobility of the enzyme in acrylamide gels in Tris-borate buffer systems at pH 8.4. Suitable stabilizing agents must be added to the system; in particular, thioglycollate is substituted for mercaptoethanol. The active pools from these electrophoresis runs exhibit two bands when examined on analytical gel electrophoresis and are about 200-fold purified relative to starting crude extracts (21).

3. *Digestion of PCA-Peptides and PCA-Proteins*

PCA-Peptides. Small peptides suspected of having terminal PCA can be conveniently studied with "G-200 enzyme". In the case of larger peptides, where secondary proteolysis might be a problem, it is desirable to use the more purified "A-25 enzyme". In either case paper electrophoresis at pH 2 is a simple method for testing the effectiveness of the enzyme prior to a detailed sequence study, since peptides which lose a terminal PCA residue have their cathodic mobility significantly increased by the exposure of the (penultimate) α-amino group. Ninhydrin negative peptides also become ninhydrin-positive, of course.

Typically, a frozen subpellet of enzyme extract is thawed and dissolved in a suitable volume of 0.05 M phosphate buffer containing 0.01 M mercaptoethanol and 0.001 M EDTA to yield an A_{280} between 2 and 4. The material is then dialyzed against the same buffer to remove the residual ammonium sulfate and 2-pyrrolidone (the stabilizer is also an inhibitor). Usually three successive 1 h dialyses against 250 ml buffer each are sufficient. Alternatively, the subpellet can be dissolved in a smaller volume of the same buffer and passed over a small (1.0 × 10 cm Sephadex G-25 column to remove the ammonium sulfate and the 2-pyrrolidone. The optimum concentrations of enzyme and substrate will vary with the particular peptide under study, and it is useful to make a preliminary examination of the reaction progress on a small amount of the peptide before attempting to prepare a large amount for subsequent degradation or other follow-up characterization. To this end, about 50 nmol of peptide are dissolved in 100 µl of the enzyme solution and incubated at 30°. Aliquots (20 µl each) are removed periodically over the course of 0-20 h and applied directly to electrophoresis paper to stop the reaction. The papers are then electrophoresed at pH 2.0 (8% acetic acid, 2% formic acid) and stained appropriately (Sect. II).

Besides the change in mobility of the substrate peptide, successful removal of terminal PCA residues can be observed by following the appearance of the PCA itself. At pH 4.1 PCA has a mobility (anodic) twice that of free aspartic acid. We use a 0.1 M pyridine-acetic acid buffer, pH 4.1, 300 V, 2 h and locate the PCA with the chlorine gas method.

If the digestion appears satisfactory and complete in the pilot study, another subpellet is thawed, dialyzed, and proportionally more enzyme

Table 3. Amino-terminal analysis of bovine and human fibrinogens digested with pyrrolidone carboxyl peptidase[a]

		0 h	4 h	8 h
Bovine				
	Glutamic acid (α-chain)	0.99	0.86	0.97
	Phenylalanine[b] (β-chain)	0	0.53	0.70
	Tyrosine (γ-chain)	1.00	1.00	1.00
Human				
	Alanine[c] (α-chain)	0.87	0.97	0.99
	Glycine[b] (β-chain)	0	0.92	1.06
	Tyrosine (γ-chain)	1.00	1.00	1.00

[a] The enzyme digestions were carried out by mixing 0.5 ml of fibrinogen solution ($A_{280} = 50$) with 0.5 ml of "A-25 enzyme" solution ($A_{280} = 1.8$) and incubating at 30°. Digestions were terminated by the addition of 1.0 ml of pyridine. Amino-terminal analysis was conducted by the thioacetylthioglycolate (TATG) procedure, which regenerates the free amino acid (309). The results are given in mol per mol of tyrosine, the common amino-terminus of fibrinogen γ-chains (fibrinogen is a dimeric six-chained protein, $\alpha_2\beta_2\gamma_2$, of molecular weight 340,000).

[b] Phenylalanine and glycine are known to be the penultimate amino acids following the terminal PCA in the β-chains of bovine and human fibrinogens, respectively (108). The absence of other terminal residues in the digest can be observed directly in the amino acid chromatograms shown in Figure 2.

[c] A small fraction of human α-chains lacks the terminal alanine and begins instead with the penultimate aspartic acid residue.

solution used to digest a substantial amount of the peptide for an appropriate length of time. The new species, without the terminal PCA, can be recovered by preparative electrophoresis or ion exchange chromatography on Dowex 50X2. Alternatively, the digestion can be terminated by the addition of an appropriate organic solvent in order that an end group analysis or a step-wise degradation may be undertaken directly.

PCA-Proteins. Removal of terminal PCA moieties from proteins has been less well studied. In general a sustained digestion with "A-25 preparations" of the enzyme is recommended. For enzyme protein concentrations of the order of $A_{280} = 1-2$, incubations of up to 16 h at 30° are advisable, although in some cases shorter times are sufficient (Table 3). The digestions can be stopped by the addition of an appropriate organic solvent so that end group analysis can be undertaken. If availability of material permits, control incubations of enzyme solution alone and protein alone should be run simultaneously. In our hands, "A-25 enzyme" solutions of $A_{280} = 1.0-3.0$ do not reveal significant quantities of endogenous end groups when incubated as controls. The same preparations readily remove the terminal PCA group from the β-chain of two different native fibrinogens, however, exposing the penultimate phenylalanine residue in the case of the bovine molecule and the penultimate glycine

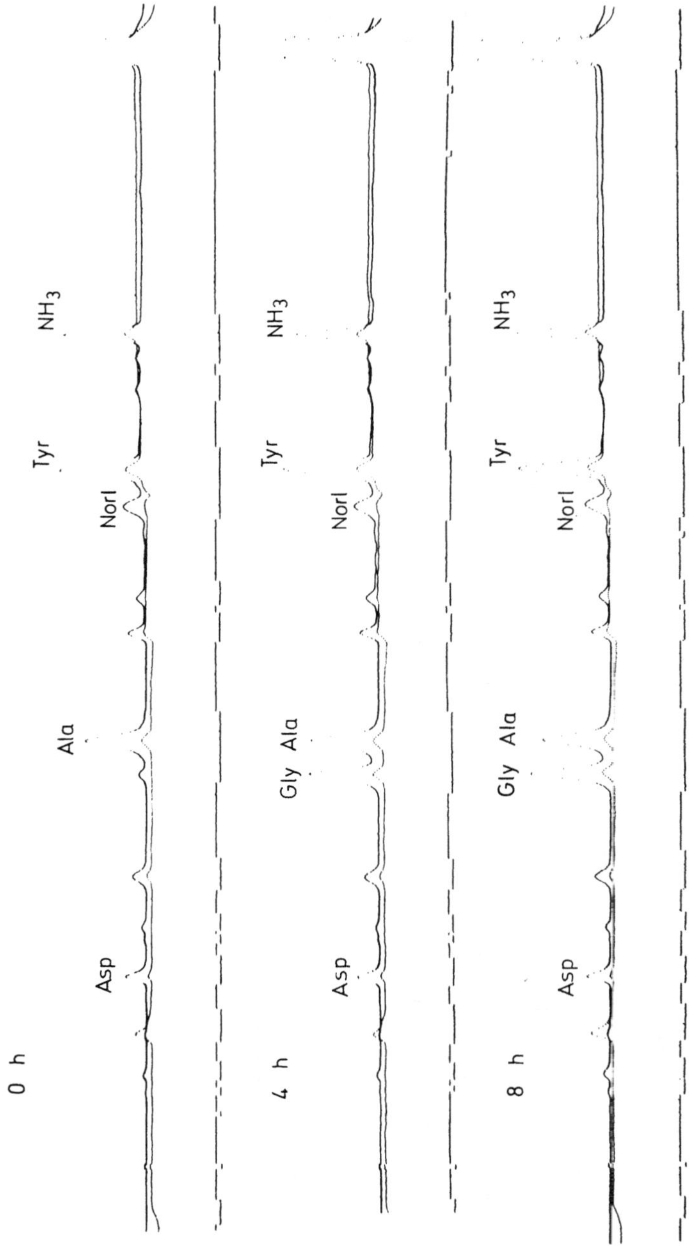

<u>Fig. 2.</u> Amino acid analysis charts of amino-terminal amino acid released by thioacetylation of human fibrinogen digested with PCA peptidase. Chromatograms were obtained with a Spinco Model 119 automatic amino acid analyzer employing a three buffer single column system. NORL norleucine internal standard. The digestion conditions and quantitation of data are presented in Table 3 (133)

residue in the human protein (Table 3). No other amino-terminals are exposed during the digestion (Fig. 2).

Although the terminal PCA in these native fibrinogens is readily removed by pyrrolidone carboxylyl peptidase, other proteins may present more difficult situations. It is generally accepted that the fibrinopeptide portions of fibrinogen molecules are on the surface of the protein, and the terminal PCA may be less accessible in other proteins. On the other hand, Mosesson (personal communication) has treated cold insoluble globulin with PCA peptidase and exposed a penultimate alanine residue, and cold insoluble globulin is an even larger protein than fibrinogen. As for smaller proteins, Szewczuk and Mulczyk (423) have reported the release of PCA from human seromucoid using an enzyme prepared from *B. subtilis*, although they did not attempt to establish any newly exposed amino-terminal residues.

4. Some Pertinent Properties of Bacterial PCA-Peptidase

Stability. Pyrrolidone carboxylyl peptidase is a sulfhydryl-dependent enzyme and is easily poisoned by iodoacetamide and organomercurials. It is necessary to keep a reducing agent present in solutions employed for isolation steps as well as in the final mixture. The active enzyme sulfhydryl group can be reversibly blocked by reaction with tetrathionate under conditions similar to those used to block other sulfhydryl proteases (21). Subsequent short term incubations with mercaptoethanol restore full activity.

The bacterial enzyme is remarkably stabilized by the presence of 0.1 M 2-pyrrolidone (IV).

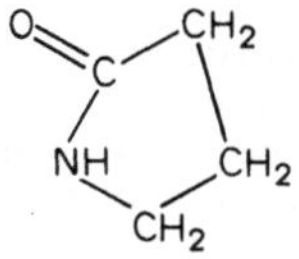

(IV)

The substance, which was selected as a potential substrate analogue, behaves as a noncompetitive, fully reversible inhibitor of the enzyme (21). It must be dialyzed away before the enzyme is used for digestion purposes.

Generally speaking, the purer the preparation of pyrrolidone carboxylyl peptidase, the more liable it is to decay during storage. Hence, "G-200 enzyme", frozen as pellets after precipitation with ammonium sulfate, is stable at -20° for several months, whereas "A-25 enzyme" stored the same way loses about half its activity in six weeks. Purer preparations decay even more rapidly. The activity of all these preparations is preserved considerably longer when storage is at -70° (119).

The instability of the enzyme at room temperatures has proven a nuisance with regard to supplying other workers with active preparations, shipments having to be made on dry ice with all the problems attendant in special handling.[2] Accordingly, we have searched for a simple method

[2] Since our original report describing this enzyme (120), we have responded to more than a hundred requests for the bacterial strain, dried bacteria, purified enzyme, and/or suitable PCA-peptides for substrates. More than half of these shipments were made on dry ice. Several of these ended unhappily when the package was allowed to sit in post offices or delivery rooms for inordinate periods.

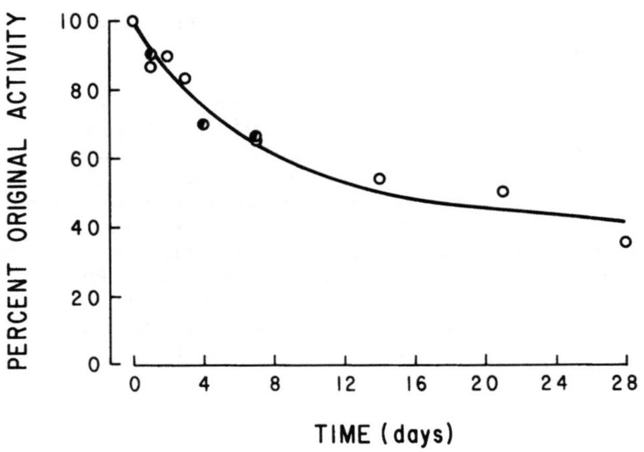

Fig. 3. Stability test of PCA peptidase ("A-25 enzyme") stored as an ammonium sulfate slurry at room temperature (T = 21-23°)

of stabilizing enzyme preparations for shipping at ambient temperatures In this regard, we now prepare ammonium sulfate slurries (containing 2-pyrrolidone) which can be mailed in an ordinary manner, since the activity is reasonably preserved at room temperatures for several weeks (Fig. 3).

Specificity. A consideration of specificity for pyrrolidone carboxylyl peptidase has two aspects. On the one hand, will it attack all or most PCA-peptides, no matter what the nature of the adjacent amino acid? Secondly, is the enzyme specific for the peptide bond linking PCA to other amino acids or is this activity only part of a much broader range of specificity? With regard to the first aspect, so far the only linkage known not to be cleaved by the enzyme is the L-pyrrolidone carboxylyl L-proline (PCA-PRO) peptide bond (443). On the other hand, the rate at which the enzyme hydrolyzes different PCA-dipeptides varies considerably. Of all the substrates tried so far, L-pyrrolidone carboxylyl L-alanine (PCA-ALA) is the most readily attacked (Fig. 4). Crude ex-

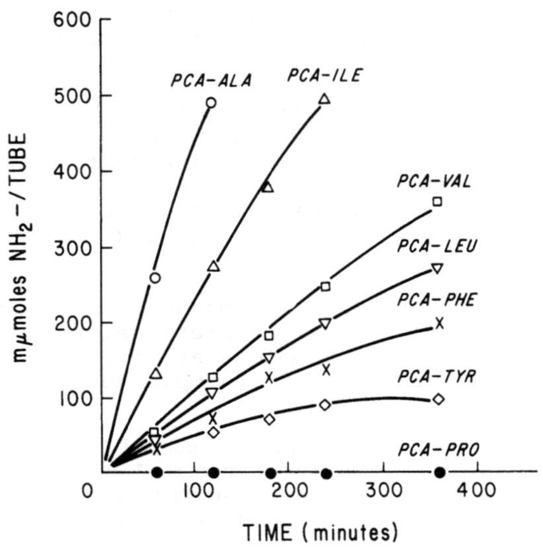

Fig. 4. Hydrolysis rates of seven L-pyrrolidone carboxylyl L-amino acid dipeptides. T = 30°; pH = 7.3. From Uliana and Doolittle (443)

Table 4. Some naturally occurring proteins and peptides with acetylated amino-terminals[a]

	Acetylated amino acid
Tobacco mosaic virus (assorted chains)	Serine
Tobacco mosaic virus (strain ORS)	Methionine
Turnip yellow mosaic virus	Methionine
Histones (assorted types)	Serine
Cytochrome c (plants)	Alanine
Cytochrome c (animals)	Glycine
Ovalbumin (assorted birds)	Glycine
Hemoglobin α-chain (carp)	Serine
Hemoglobin γ-chain (human)	Glycine
Actin (rabbit muscle)	Aspartic acid
Factor XIII (human, bovine)[b]	Serine
Apoferritin (horse)	Serine
Alcohol dehydrogenase E chain (horse)	Serine
Keratin (sheep)	Alanine
α-Crystallin (bovine)	Methionine
Muscle calcium binding protein (fish)	Alanine
Carbonic anhydrase C (human, bovine)	Serine
Encephalitogenic protein (human, bovine)	Alanine
α-Melanotropin (assorted mammals)	Serine

[a]Taken from Dayhoff's Atlas (108, 109).
[b]From References (313) and (424).

tracts exhibit the same order of relative activity, dispelling any notion that one enzyme of a family has been inadvertantly purified by using PCA-ALA as a test substrate during the isolation procedure.

IV. Acylated Peptides and Proteins

The frequency of acylated amino-terminals is comparable to that of PCA-blocked peptides and proteins. Most of the problem for amino acid sequence studies resides with acetylated amino-terminals (Table 4), even though the formylation of amino groups is widespread in the biological world. The reasons for this are two-fold: firstly, although many prokaryotes initiate polypeptide synthesis with N-formyl methionine, these same organisms usually have an enzyme for removing the formyl group from the nascent chain (2). Secondly, the formyl group is much more readily removed from amino groups under mild conditions than is the acetyl group, being especially vulnerable to mild acid hydrolysis (166). Either of these factors may account for the fractional occurrence of formylated amino-terminals observed in some proteins (159, 239).

As for the acetylated amino-terminal, its distribution with regard to the particular amino-terminal amino acid is decidedly nonrandom, serine being the acetylated residue in almost half the known cases. Alanine, glycine and methionine have also been reported for several different proteins, but the existence of other acetylated amino acids at the amino-terminus of naturally occurring peptides and proteins is rare indeed (Table 4).

A. Chemical Identification Procedures

1. *Hydrazinolysis*

The most popular approach for the identification of an acyl group at the amino-terminus of a peptide involves the hydrazinolysis procedure introduced by Akabori and co-workers (8). If the peptide is not too large, the acetyl or formyl hydrazide can usually be identified on paper electrophoresis, since they are more electronegative than the amino acid hydrazides. However, in the case of proteins, the sheer bulk of amino acid hydrazides interferes with such simple procedures. Phillips (344) treated the hydrazide mixture obtained upon hydrazinolysis of a protein with fluorodinitrobenzene at low pH and then extracted the dinitrophenylated acyl hydrazides with an organic solvent. Schmer and Kriel (395) have described a similar procedure for identifying the fluorescent DNS-derivatives of acetyl or formyl hydrazides. As in the Phillips approach, derivatization with DNS-chloride is performed at low pH in order to preclude the DNS reaction with α-amino groups; the DNS-acetyl and formyl hydrazides are readily extracted with a suitable organic solvent from the bulk of positively charged DNS-hydrazides (Fig. 5) and can be easily identified using the popular polyamide thin layer system described by Woods and Wang (471) (Fig. 6).

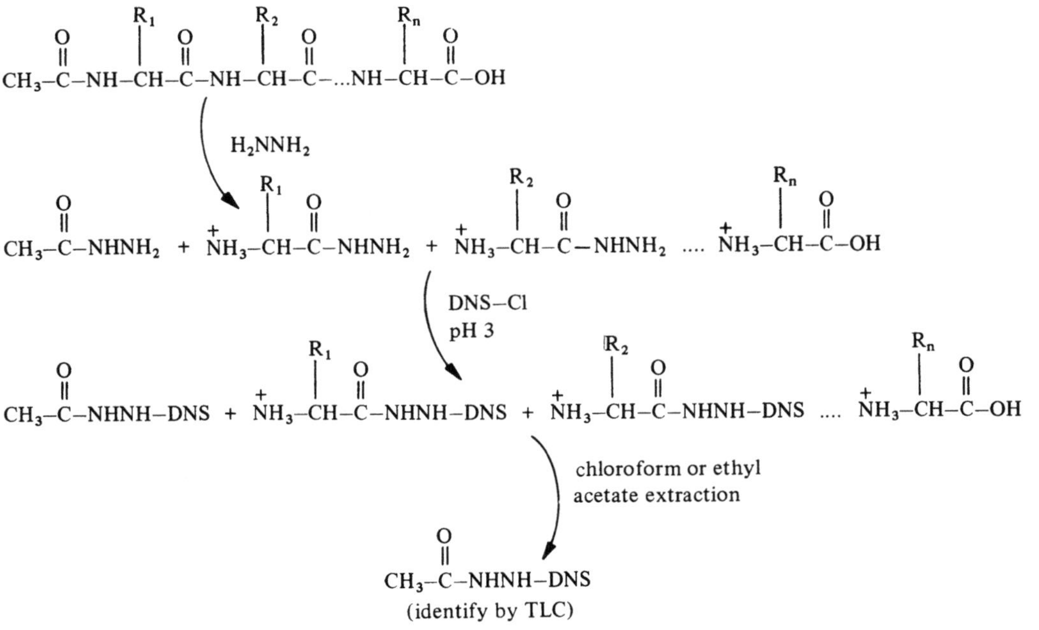

Fig. 5. Procedural outline for the hydrazinolysis of a peptide followed by treatment with DNS-Cl at low pH for identification of acyl hydrazides [adapted from (8, 344 and 395)]

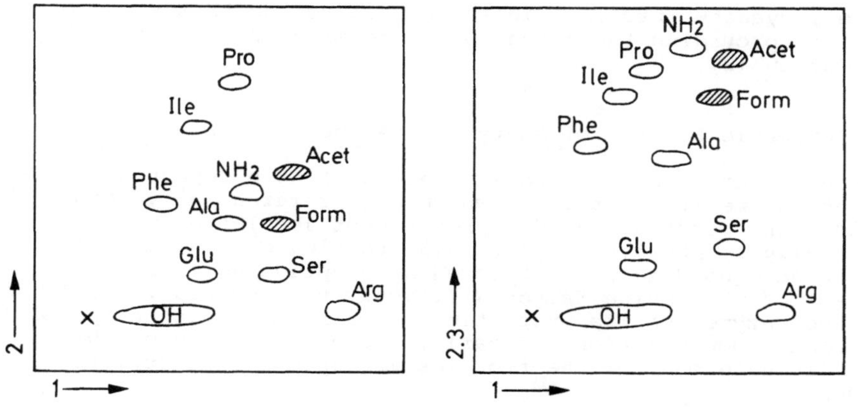

Fig. 6. Positions of acetyl-DNS-hydrazide (Acet) and formyl-DNS-hydrazide (Form) on polyamide sheets using the standard 1:2:3 solvent system of Woods and Wang (471). The relative positions of some DNS-amino acids (not hydrazides) are also shown. From Takagi and Doolittle (424)

2. Methanolysis

In principle, methanolysis is an ideal approach to the identification of acyl blocking groups, and a micromethod designed for determining N-acetyl groups in mucopolysaccharides can be applied directly to peptides (269). Depending on the nature of the blocking group, the corresponding methyl ester will be formed (methyl formate, methyl acetate, methyl propionate, etc.) and these can be isolated by microdistillation and quantified colorimetrically as the corresponding hydroxamates (269). In practice, the method has a number of drawbacks. First of all, it requires substantial amounts of material, and it is ill-advised to trust negative results achieved with less than 1.0 μmol of sample. Secondly, great care must be taken to rid samples of contaminating substances – especially acetic acid (or acetate) which is frequently employed in buffers used in the isolation of the peptides.

3. Miscellaneous Considerations

In addition to the various hydrazinolysis procedures and methanolysis, acyl blocking groups can be identified by gas chromatography after acid hydrolysis of the peptide or protein (455). The method demands considerable amounts of material, however, since the released acid (acetic acid in the case of acetylated amino groups) must be extracted from the hydrolysate with a suitable organic solvent before injection into the gas chromatograph. As in the other method discussed, great care must be taken to eliminate contaminating acetic acid or acetate, and there is, of course, no distinction made between acetic acid derived from a blocked amino-terminal and acetylated (N or O) groups elsewhere in the peptide or protein.

Nor is mass spectrometry particularly suited for the determination of acyl blocking groups, although the acetyl fragment ion with m/e = 43 has been ascertained in an acetylated peptide (68). One problem is that fragments of similar mass/charge ratio occur frequently in peptides and proteins as a result of the breakdown of amino acid sidechains.

In one regard, Nature has been kind with regard to acyl blocking groups in that methionine may be the terminal amino acid (Table 4). In these cases, cyanogen bromide digestion will release N-acyl homoserine, as well as providing the parent amino-terminal peptide for step-wise degradation (27).

B. Enzymatic Removal of Acetylated Amino-Terminal Residues

Recently Narita and his colleagues (315) have reported the isolation of an enzyme from rat liver which can preferentially remove the acetylated amino-terminal residue from a peptide. The enzyme was used successfully by Nakamura et al. (313) to cleave the N-acetyl seryl-glutamic acid bond in an acetylated nonapeptide derived from an enzymic digest of the bovine factor XIII activation peptide. A full description of the enzyme and its properties is currently being prepared (K. Narita personal communication). Clearly, the availability of this enzyme ought to eliminate many of the troublesome aspects of acetylated α-amino groups.

V. Concluding Remarks

The secret behind the successful identification of blocked amino terminals is versatility. The investigator ought not to prejudge the situation, and the possibility of having to employ a variety of procedures must not be minimized. This would include methods which are not in the everyday arsenal of most sequencers. In particular, the mass spectrometer ought to be used more routinely for the identification of PCA residues in small peptides, although that instrument is not so well suited for identifying acetylated amino-terminals. Enzymatic procedures are highly recommended, of course, and enzymes for removing both PCA and acetylated amino acids are now available (119, 442). These must be used with care, however, and the receipt of a gift enzyme package is never a guarantee of success. At one stage we made a systematic inquiry of persons to whom we had shipped PCA peptidase in an effort to find how successful the approach had been. Apart from those who had never gotten around to using the preparation, the group was divided about equally between successes and failures. Of the failures, about half of these were because the blocking group was evidently not PCA. In some other cases, the enzyme was not used in the recommended way, but in a few instances – particularly in the case of some immunoglobulin heavy chains – negative results have been obtained under well-defined conditions. It is hoped that the reasons for these uncharacteristic failures can eventually be uncovered. In the meantime, a number of alternative procedures are available and ought to be called upon when necessary.

Acknowledgments. I am grateful to my assistant, Mrs. Marcia Riley, for her help over the years in preparing large amounts of PCA-peptidase for shipment to other workers. This work was supported by grants from the U.S. Public Health Service.

Use of Antibody in the Study of Protein Structure

M. REICHLIN

I. Introduction

An exposition of the characteristic features of the antigen-antibody (Ag-Ab) reaction is essential to a discussion of the use of specific antibody for the study of the structure of proteins. The reaction can be viewed as occurring between a specific chemical grouping on the protein surface (antigenic determinant) and a complementary chemical grouping (Ag binding site or Ab combining site) found on the antibody molecules. These are serum proteins of slow electrophoretic mobility found in all vertebrate species. Antibodies comprise a family of proteins called the immunoglobins which have a tetrameric polypeptide chain structure with two pairs of chains of unequal size (so-called light and heavy chains) held together by both noncovalent forces and disulfide bonds. The molecule is composed of two symmetrical halves each containing an antibody combining site whose complementary structure is formed by the three-dimensional confluence of three to four hypervariable regions found in the amino-terminal halves of both the light and heavy chains. Extensive discussions of both the general structure of the immunoglobins and the particular structure of the Ab combining site can be found in recent reviews (83, 316).

Some general principles regarding the molecular location of the antigenic determinants in globular protein antigens are becoming manifest and an extensive discussion of such data can be found in a recent review (368). What emerges from such discussions is that the antigenic determinants in proteins are found in surface locations which are in general where that family of proteins has undergone evolutionary variation, and specifically where there are differences between the amino acid sequence of the immunizing protein and that same protein in the immunized animal. For example, the antigenic determinants of human hemoglobin reactive with antisera prepared in rabbits are found in regions of amino acid sequence where rabbit and human hemoglobins are different. It is inherent in that statement (and an experimental fact) that rabbit hemoglobin is unreactive with antibodies produced in rabbits to human hemoglobin in later sections of this chapter, the nature, number, and location of these determinants in human hemoglobin will be discussed. Other aspects of the antigenicity of human hemoglobin will be discussed to illustrate other points relevant to the theme of this chapter.

The reaction of a protein antigen with antibody is reversible and can be viewed as a special class of protein-protein interaction. The chemical bonds involved are noncovalent in nature, such as hydrophobic, hydrogen, and ionic. The affinity constants for such reactions have been measured and vary widely, occurring in the range from 10^5 to 10^{11} M^{-1}. It is clear that the physicochemical techniques used to measure such reactions must be of appropriate sensitivity and precision not only to detect these reactions but also to detect small binding dif-

Table 1. Quantitative techniques for measuring the Ag-Ab reaction

Type of assay	Ab concentration in mg per ml
Primary binding methods	
Ultracentrifugation	
Schlieren optics	10^{0}
Absorption optics	10^{-2}
Farr technique radioimmunoassay	10^{-6}
Fluorescence quenching	10^{-2}
Methods dependent on secondary manifestations of Ag-Ab reaction	
Quantitative precipitin analysis	10^{-1}
Passive hemagglutination (and inhibition methods)	10^{-5}
Enzyme inhibition	$10^{-6} \rightarrow 10^{-5}$
Methods dependent on tertiary manifestations of Ag-Ab reactions	
Complement fixation	
Macro technique	10^{-3}
Micro technique	10^{-5}

ferences between closely related antigens. It is these matters that are the subject of this chapter and they can be amplified by posing two questions that relate directly to the power of resolution of immunochemical methods to discriminate between two closely related protein antigens in their reaction with specific antibody.

The questions are the following: (a) can antibody detect a single amir acid difference between two proteins? (b) do conformational changes ir proteins affect antigenicity?. The answer to both these questions is a qualified yes in that many but not all single amino acid changes in proteins can be detected by specific antibodies and similarly many but not all conformational changes in proteins can be detected by specific antibody. Indeed, an analysis of which single amino acid changes in a homologous set of proteins can affect antigenicity has been a major tool for the definition of which portions of the molecular surface of proteins are antigenic. Before entering into any detailed discussion of specific examples of antigenicity, a discussion of the techniques for measuring the Ag-Ab reaction is in order.

II. Measurement of the Ag-Ab Reaction

Such techniques can be classified by whether they (a) reflect some prc perty which is only a function of the binding of Ag to Ab, so-called primary methods, or (b) whether they reflect some secondary or tertiar manifestation of the Ag-Ab reaction which occurs subsequent to the primary binding step.

Table 1 is a summary of the more important quantitative methods that have been used in the study of the Ag-Ab reaction. Listed in the table are the techniques and their approximate sensitivity for the detection of specific antibody. These lower limits must be taken as approximate since there is great heterogeneity in affinity among antibodies and the sensitivity of detection obviously depends on affinity at a given concentration. A few words are in order about the information that can be extracted from measurements involving each of these methods.

For quantitative analytical work the primary binding methods are essential for the study of the Ag-Ab reaction for the reason that by far the most information can be extracted from such measurements. For a thorough characterization of the Ag-Ab reaction one would like to know the basic chemical data necesssary for the evaluation of any chemical reaction: the stoichiometry of the reactants, the nature of the products as well as the determination of the kinetic and the affinity constants of the reactions involved. Such data can only be obtained with purified reactants in which the concentration of all reactants and the resultant complexes can be accurately determined in kinetic experiments and at equilibrium. In experiments involving solution methods, the aggregation and phase separation which results from multivalent protein antigens reacting with bivalent antibody must be circumvented. This is done by converting bivalent immunoglobulin (MW 150,000) into monovalent Fab fragments (MW 50,000) by proteolysis with pepsin or papain (368). Such preparations can then be used in ultracentrifugation and fluorescence quenching experiments. Ultracentrifugation can be used to determine reaction stoichiometry (467). This technique is appropriate for the determination of equilibrium constants when there is significant dissociation of complexes at the concentrations of reactants which can be measured by this technique. Since the molar concentration of reactants in this method is in the range of 10^{-5} M (with Schlieren optics) and 10^{-7} M (with absorption optics) and the equilibrium dissociation constants for protein-antiprotein reactions are for the most part smaller than 10^{-8} L/M (Reichlin, unpublished experiments), this method will not be very useful for the determination of equilibrium constants for many protein-antiprotein reactions.

Fluorescence quenching is a method that can be used when there is a chromophore in the protein antigen which can quench the tryptophan fluorescence of antibody. Heme protein antigens such as hemoglobin and cytochrome c are ideally suited for such assays, and this method has been extensively used for the determination of reaction stoichiometry in these systems (327, 328). The concentration of reactants in these assays is similar to that in ultracentrifugation performed with absorption optics and precludes the determination of equilibrium constants for systems with K_{eq} greater than 10^{7} M^{-1}. At the concentrations employed, the titrations are stoichiometric with little or no free antigen present when incremental additions are made to antibody solutions.

The Farr technique, which is a primary binding method, is a versatile, sensitive, and accurate means for studying the Ag-Ab reaction. Purified antigen is radiolabeled to the desired specific activity with ^{131}I or ^{125}I and after an appropriate length of incubation, bound antigen can be separated from free antigen by chromatoelectrophoresis (40), differential solubility in ammonium sulfate (148), or by the double antibody method (364). This technique and its variants will probably supplant all others for the study of antigenic specificity because of the simplicity of the quantitative measurements, the inherently great sensitivity of the method for distinguishing subtle changes in protein structure, and the broadness of application.

Precipitin analysis is of great historic importance as it was the first quantitative method for studying the Ag-Ab reaction (181). It is still of importance in determining the quantity of antibody in serum although immunoabsorbent methods (wherein antigen is attached to an insoluble support) are better, since the latter measure nonprecipitating as well as precipitating antibody. With certain small antigens such as cytochrome c, the majority of antibody is of the nonprecipitating variety (365). The precipitin technique is also relatively insensitive, which limits its usefulness. Passive hemagglutination and enzyme inhibition are both very sensitive techniques and have been used to study relatedness of protein antigens (241), as well as conformational changes in proteins (382) with great facility and exactitude.

Finally, complement fixations has been used extensively to study antigenic structure in proteins. The method is very sensitive and precise and has proved a powerful immunological tool in the study of protein structure. It is most useful with rabbit antisera since goat, horse, and guinea pig antisera have a very significant proportion of antibody with heavy chain structures that do not interact with complement when aggregated by antigen. The great majority of antibody to protein antigens which is found in hyperimmune rabbit sera fixes complement predictably and efficiently.

Once a quantitative method of studying the antigen-antibody reaction is in hand, antigenic structure can be studied in a number of ways. Firstly, the native unmodified antigen can be directly compared to closely related antigens modified by amino acid substitution or group-specific chemical modification in their reactions with specific antibody. Antigens with their determinants modified will react less well than native antigen with antibody. Modifications outside the molecular limits of a determinant will not, in general, affect antigenicity unless they induce an important distant conformational change which affects the structure of the determinant. This latter problem will be discussed in the next section.

Historically, the most definitive method for determining antigenic structure has been the technique of haptene inhibition. With carbohydrate antigens, the polymeric polysaccharide can be degraded to oligosaccharides which generally are capable of binding to antibody. In competition experiments between polymer and oligosaccharides for antibody sites, one can deduce the structural requirements for antigenicity by determining the most effective size, monosaccharide sequence, and chemical linkage that inhibit the reaction between polysaccharide and antibody. Experiments such as these led to the first estimates of the size and shape of antibody combining sites (223). Unfortunately, with globular proteins this method has been of only limited value because antigenicity is largely lost when the specific three-dimensional structure of the antigen is destroyed by specific proteolysis. A fraction of the determinants in globular proteins can be studied in this way and when peptides bind to antibody their activity can be assessed by their ability to inhibit the reaction of the native protein antigen with antibody. The antigenicity of small peptide hormones can be quite successfully studied in this way, presumably because there is little three-dimensional structure in such peptides and precise molecular requirements for antigenicity can be defined in terms of unique primary sequences.

This brings the discussion of the role of conformation in studying antigenicity.

III. The Effect of Conformational Changes on the Antigenicity of Protei

A great deal has been written on the effect of conformational changes (the antigenicity of proteins. Detailed descriptions of such effects ca be found in several recent reviews (38, 86, 368, 399). One can in general say that sequence changes in proteins may affect antigenicity by either inducing conformational changes or by changing the chemistry within the molecular domain of the determinant without significant distant effects on conformation in the protein. Although there are few situations in which one could be absolutely sure that distant conformational effects are not induced by single residue changes in proteins, certain generalizations have been made which seem representative of the facts (368). Within homologous gene families there is overall preservation of the three-dimensional structure of proteins and while conformational differences are occasionally found among certain proteins of the same gene family, such examples are not common. Data are available for the myoglobins, cytochromes, lysozymes, and hemoglobins where various types of measurements of conformation, including X-ray diffraction, reveal overall preservation of the three-dimensional structure within a homologous set of proteins.

In addition, single amino acid changes are not often associated with measurable conformational changes (be they induced by mutation or chemical modification) and when conformational changes due to single amino acid changes do occur they only infrequently affect antigenicity. If antibody is utilized for the purpose of detecting sequence change per se, it makes little difference whether sequence changes affect antigenicity via local chemical changes in the determinant or by long-distance conformational changes. However, in any specific case it is sometimes possible to decide if a single sequence change in a protein induces its antigenic effect by a local structural change (the usual situation) or by distant conformational effects.

What is perhaps potentially most exciting from the standpoint of the protein chemist are the potential applications of antibody to study very specific aspects of protein structure. Data exist which suggest that antibody can be used in the following ways:

1. As a reagent which can sense specific conformational states of the same protein.
2. As a conformational probe for specific molecular regions of proteins.
3. As a biochemical means of identifying specific amino acid sequence within a given protein.
4. As a molecular tool for assessing the overall rate of protein evolution with a given gene family.

These applications will be discussed sequentially.

Specific antibody is capable of distinguishing the conformational differences that exist between myoglobin and its apoprotein (101, 361), oxy- and deoxyhemoglobin (362), oxidized and reduced cytochrome c (363), and various conformational states of the enzyme staphylococcal nuclease controlled by calcium ion and deoxythymidine nucleotide (158). These are only a fraction of the conformational differences that can be detected by antibody, but are listed because the conformational changes in all cases are reversible and controlled by the addition of small molecules so that in these cases the altered antigenicity can be associated with reversible conformational changes. There is therefore no doubt that antibody can sense conformational changes in antigens but it is equally clear that not all conformational changes in protein antigens are detectable by specific antibody. From the practical standpoint the problem arises as to the mode of interpretation of an antigenic change when studying a particular protein. As either a sequence change or a conformational change can effect antigenicity, it is impor-

Table 2. Conformational changes affecting the antigenicity of hemoglobin

1. The difference between deoxyhemoglobin and liganded hemoglobin.
2. The difference between globin and hemoglobin
3. The conformational effects arising from heterotropic subunit interactions
4. The difference between carboxypeptidase A modified hemoglobin and normal hemoglobin.
5. The difference between hemoglobin bound to haptoglobin and free hemoglobin.

A discussion of the first four of these effects can be found in reference (368) while detailed information about antigenic changes in hemoglobin when bound to haptoglobin can be found in reference (213).

tant when studying a given protein to ensure that factors known to affect a specific protein's conformation be controlled so that a change in antigenicity can be ascribed to a sequence change and not to an unrecognized conformational change unrelated to sequence changes per se. For example, in Table 2 are listed a number of recognized conformational changes in human hemoglobin which affect the quantitative measure of antigenicity. Thus ligand binding, apoprotein formation, subunit interactions, loss of the carboxyl terminal histidine from the β chain and the binding to haptoglobin all lead to measurable conformational changes which affect antigenicity. It is apparent then that in the antigenic comparison of any two proteins they must be studied under exactly the same microenvironmental conditions.

The use of specific antibody as a molecular probe has been most elegantly applied to the study of a conformational transition in the enzyme staphylococcal nuclease. A fraction of antibody which reacts with a limited region of the carboxyl-terminal sequence (99-126) of the enzyme can be isolated by immunoabsorbent methods (382). This antibody fraction binds with unitary stoichiometry to enzyme (one mol of Ab sites per mol of enzyme) and the complex is inactive. Peptide 99-149 which has no ordered structure in solution (while the same region in the protein is highly ordered), nonetheless, in high molar excess, can complete with native enzyme for binding to antibody related to sequence positions 99-126. Quantitative assessment of this competition led to the determination of an equilibrium constant for the proposed equilibrium between the native and unfolded forms of the peptide, as it was assumed that only the form of the peptide having the conformation of native enzyme could bind antibody. This interpretation has been buttressed by recent experiments of a reciprocal nature in which antibodies to the peptide 99-149 were studied (160). These antibodies had the reciprocal property of binding well to the peptide and not at all to native enzyme. Studies such as these point to the exciting possibility that antibodies sensitive to specific regional conformational changes in proteins can be used as molecular probes for studying various parts of proteins independently.

IV. Detection of Sequence Changes in Proteins with Antibodies

If one takes into account the difficulties that might arise due to conformational changes in protein antigens, immunochemical techniques have unsurpassed sensitivity in detecting and in favorable circumstances

Table 3. Protein antigens in which single amino acid substitutions have been demonstrated to affect antigenicity

Human Hb (368)	Species in which antibodies prepared
Human cytochrome c (326)	Rabbit, goat
E.coli alkaline phosphatase (91)	Rabbit
Egg white lysozyme (350, 351a)	Rabbit
Sperm whale myoglobin (26)	Goat

[a]In this case single substitutions have not been studied per se but proteins differing by a small fixed number of residues (3-6) have been distinguished from one another by quantitative microcomplement fixation.

identifying amino acid sequence changes in proteins. Thus it is well known that many single amino acid changes can be detected by immunochemical methods and that in certain circumstances antibodies can be isolated from sera which are apparently uniquely reactive for sites controlled by specific single amino acid residues. Table 3 lists a number of protein antigens in which single amino acid changes have had sufficient impact on antigenicity to be detectable by specific antibody. Except for the myoglobin system all of the single amino acid changes were naturally occurring mutant proteins or molecular representative of closely related species. In the case of myoglobin the single amino acid changes were introduced by group-specific chemical modification. It is apparent that quantitative immunochemical methods can detect single amino acid changes in protein antigens. Indeed, single changes can make such decisive contribution to antigenic specificity that the binding difference between the homologous antigen and the mutant protein is sufficiently large that the mutant protein does not bind to the antibody population related to the antigenic determinant affected by the mutation. Two such cases have been studied in detail (328). The most systematic study of the effect of single amino acid changes on antigenicity has been in the human hemoglobin system. In this system one can ask and tentatively answer most of the questions posed about the ability of antibody to detect sequence changes in proteins.

Firstly, what fraction of one amino acid mutant can be distinguished from Hb A_1 by specific antibody. Using the C-fixation method and rabbit anti-A_1 Hb sera, it has been shown that about one-third of the variant hemoglobins studied were antigenically deficient to Hb A_1 (366). If all the variants were studied with specific antisera to the isolated chains of hemoglobin (anti-α and anti-β), the fraction of variants distinguishable from A_1 Hb rose to 57%. Finally, although all the variants have not been studied in this way, immunization with mutant hemoglobins gave rise to sera which in several cases demonstrated specificity for the mutant hemoglobin with mutant hemoglobins which could not be distinguished from A_1 Hb with anti-A_1 sera. Thus of the mutant hemoglobins studied, the fraction giving rise to specific antibodies in rabbits rose to 84% (367). These data are tabulated in Table 4. Therefore, in the hemoglobin system if all variants are examined for their ability to elicit specific antibodies in rabbits which distinguish them from Hb A_1 the great majority give rise to such antibodies. These antimutant sera are clearly the most sensitive immunological reagents for detecting sequence changes in proteins. The interpretation of these data un-

Table 4

Fraction of variants consistently distinguishable from A_1 Hb	Mode of comparing variant to A_1
9/28 (32%)	C-fixation with anti-A_1 sera
16/28 (57%)	C-fixation with anti-chain serum
11/13 (84%)	C-fixation with anti-variant serum

derlies any understanding of how general these findings might be. One can ask what determines which mutants are detectable by anti-A_1 sera.

These data can be correlated with the sequence similarities and differences that exist between human and rabbit hemoglobins. As has been pointed out previously, the specificity differences revealed by the comparison of mutant hemoglobins to A_1 can be related to the sequence differences between rabbit and human hemoglobins.

As rabbit hemoglobin did not react with these sera, it was hypothesized that these specificity differences must relate to the sequence differences between rabbit and human hemoglobin. Comparison of the hemoglobin sequences of these two species revealed them to be nonrandom, and 90% of the differences could be segregated into one-half the total sequence Of interest was that the distribution of the variation within the alpha and beta chains was very similar. These regions were designated "variable" and the remainder "conservative". As all the distinguishable mutants were located in the variable regions and all the indistinguishable mutants were localized in the conservative regions, it was postulated that within these variable sequences were to be found the antigenic determinants of human hemoglobin.

Figure 1 is a previously published diagram (368) in which the validity of the original hypothesis has been refined by additional data. The

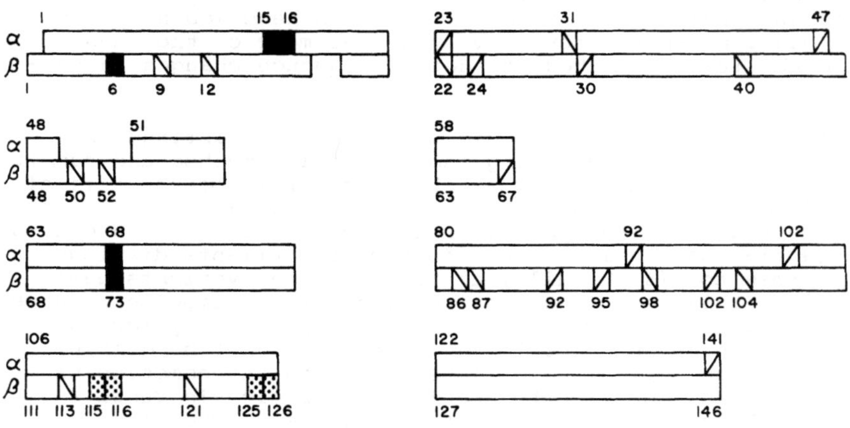

Fig. 1. Mutant Hb S antigenically distinguishable (filled box) and indistinguishable (diagonal line) from A_1 are related to sequence differences and identities between rabbit and human Hb S_1, respectively. In addition to the data listed here Hb Leiden β^6 or ^{7} glu→deletion is a distinguishable mutant while Hb Abruzzo β^{143} His→Arg is an indistinguishable mutant from A_1

left-hand side of the diagram contains the variable regions and the right-hand side the conservative regions. Each sequence position on the diagram is indicated by a box of equal length. A diagonal within a box refers to an indistinguishable mutant and a filled box to a distinguishable mutant. Dots within a box mean that one or more of the indicated sequence positions is involved in the same determinant but the data are not decisive on which position contains the responsible residue(s). It is seen that none of the evaluated positions in the conservative regions is involved in antigenic determinants; all positions involved in antigenic determinants are in the variable regions. There are also positions in the variable regions that are not involved in antigenic determinants, and these demarcate what are likely to be the molecular limits of the determinants. For example, in the β variable region 1-21, the distinguishable mutants are S (β^6 Glu→Val), C (β^6 Glu→Lys), and Leiden ($\beta^{6 \text{ or } 7}$ Glu→deletion) located near the N-terminus of the beta chain. Positions 9, 12, 22, 50, 86, and 87 in the beta chain are not involved in an antigenic determinant since Lepore Hb (which differs from A_1 at those positions) is indistinguishable from A_1 with rabbit anti-A_1 serum. Thus, the determinant at the N-terminus of the beta chain is to be found within the sequence 1-8. Positions 115, 116, 125, and 126 containing dots are so designated as these positions represent the carboxyl-terminal differences between A_1 and A_2 Hbs. One or more of these residues influences the structure of an antigenic region.

A number of points might be made about these data. Most importantly, all the mutants that are thought to be involved in antigenic determinants are surface residues. Of the mutants that are not involved in determinants, about equal numbers of externally and internally oriented side chains exist. This is further emphasized when one looks at the immunogenicity of hemoglobin mutants in rabbits (367). Specific antibodies can be produced to every mutant hemoglobin thus far tested if the mutant involves a surface residue. When the mutation involves an internal residue, mutant-specific antisera are usually not obtained. It is true that internal substitutions can occasionally affect antigenicity, but the quantitative effects are small. For example, in the lysozyme immune system, positions 40, 55, and 91 (all internal residues) contribute to an antigenic structure that distinguishes the chicken from the bobwhite quail enzyme (349). In the hemoglobin immune system, anti A_1 serum cannot detect antigenic differences between internal mutants and A_1, but an occasional serum prepared against the mutant can exhibit some specificity for the mutant hemoglobin compared to A_1 (367).

Another approach relating antigenicity to single amino acid changes in hemoglobin has involved immunization of rabbits with one amino acid mutant from both variable and conservative regions. What occurs is that all mutants from the variable region not only give rise to mutant-specific antisera but the quantitative mutant A_1 difference with antimutant serum is larger than the A_1 mutant difference with anti-A_1 serum. Moreover, either precipitin tests or C-fixation tests detect such differences when using the antimutant sera, whereas, A_1 mutant differences with anti-A sera are regularly demonstrable only by C-fixation (366, 367). Such increased sensitivity of the antimutant sera compared to anti-A_1 sera may be due either to a large fraction of antibody directed to that region in antimutant sera or a larger binding energy difference between the mutant and A_1 reacting with antimutant serum compared to the comparable parameters with anti-A_1 serum. This larger discrimination correlated in every case with a larger sequence difference between the mutant and rabbit hemoglobin compared to the number of sequence differences between A_1 and rabbit hemoglobin. These studies all exhibit a

Table 5. Correlations of the ability to discriminate human hemoglobins A_1 and S with sequence differences between immunogen and hemoglobin of injected specie[a]

Serum	C-fixation	Precipitin test	All-or-none antibody	Sequences differences
Rabbit anti-A_1	+	-	-	βA_1-β^{rabbit} (83)
Rabbit anti-S	+	+	-	βS-β^{rabbit} (368)
Goat anti-A_1	n.d.	+	-	βA_1-β^{goat} (368)
Goat anti-S	n.d.	+	+	βS-β^{goat} (467)

[a]All-or-none antibody is defined as antibody that does not bind to the heterologous antigen in fluorescence quenching experiments. The sequence differences listed are only those within the first ten amino acids of the N-termini of the beta chain. n.d., not determined (368).

positive correlation between the number of sequence differences between the immunogen and the hemoglobin of immunized species and the discriminatory power of the antiserum for the mutant hemoglobin and A_1.

Immunization of rabbits with mutant hemoglobin from the conservative regions of hemoglobin gave rise to sera showing variable behavior. Anti-sera to two internal substitutions (β^{67} Val→Glu and β^{24} Gly→Arg) usually failed to discriminate the mutant hemoglobin from A_1. Likewise a substitution in the conservative region involving loss of a side chain (α^{47} Asp→Gly) was immunogenically silent. Two external substitutions were variably immunogenic, and with the substitution β^{102} Ser→Arg a number of sera arose that distinguished the mutant from A_1 regularly by precipitin tests but not by C-fixation. By contrast, 28 of 29 sera resulting from immunization with nine different mutants from the variable regions clearly distinguished the mutant from A_1 hemoglobin by both precipitin and C-fixation tests (367). Although much more work is required on mutants from the conservative regions, the early data suggest decreased immunogenicity of variants compared to those from variable regions with the effect being quantitatively greater with internal than with external mutants.

Such data provide a tentative explanation for the restricted number of populations of antibodies arising in rabbits in response to human hemoglobin. It is that only a fraction of the surface of the protein is immunopotent and more specifically the immunopotent regions are those where there exists a concentration of sequence differences between human and rabbit hemoglobins.

Finally, a detailed analysis of the specificities arising from immunization of goats and rabbits with two hemoglobins (A_1 and S) differing by a single amino acid gives some insight into the relationships of sequence differences between immunogen and that protein in the immunized specie and binding differences between the two hemoglobins. Table 5 lists data which bear on this point in the case of human hemoglobins A_1 and S in reactions with their antisera prepared in rabbits and goats Rabbit anti-A_1 detects a difference in C-fixation reactions between A_1 and S but with such sera no precipitin difference is demonstrable between the two proteins. Moreover, Hb S can absorb out all the bindable antibody at comparable concentration of the amount of A_1 required for absorption. The antibody, therefore, that distinguishes S from A_1 in rabbit anti-A_1 serum binds A_1 only slightly better than S and does so only at the high dilutions characteristic of C-fixation reactions. On the other hand goat anti-S serum contains antibodies which will bind to HbS and not to A_1 at any A_1 concentration. We have called such fractions "all-or-none" antibodies. Rabbit anti-S and goat anti-A_1 seem to have a comparable degree of specificity in that both exhibit precipitive differences with the two antigens but in both cases extensive absorption with heterologous antigens removes all the antibody from the sera. It is reasonable to think that this increasing specificity is due to a graded difference in binding energy between the two antigens and the antibodies sensitive to the structure difference between the two antigens. The difference in binding energy of Hb S and A_1 reacting with goat anti-S is unmeasurably large, whereas the difference in binding energy of Hb A_1 and S reacting with rabbit anti-A_1 is quite small. It would be expected that the binding differences for the two cases (rabbit anti-S and goat anti-A_1) would show some intermediate value for the difference. Correlated with this increasing specificity is the number of sequence differences between the immunogen and the hemoglobin of the immunized specie in the relevant portion of the β chain sequence. Thus, there are two differences between βA_1 and β^{rabbit} and four between βS and β^{goat} in the first ten amino acids associated with a weak and a strong specificity, respectively. Thus a hierarchy of specificity probably reflecting a hierarchy of binding differences seems linked to the quantity of sequence differences that reflects "foreignness".

It is clear that antibodies such as the goat antival fraction can be used as specific reagents to identify the N-terminal sequence of human hemoglobin S. Moreover this could be done in situations where the Hb S is contaminated by a protein differing by a single amino acid (Hb A_1) where the Hb S is present in nanogram quantities and the contaminating protein is present in 10,000 times the concentration of the Hb S.

Such immunological reagents represent the ultimate in sensitivity and specificity for the detection of single-sequence changes in proteins and are only beginning to be exploited in biochemical application.

V. Use of Antibody in the Study of Protein Evolution

One of the most useful applications of specific antibody is the ability to compare different proteins of a homologous family and obtain some estimate of their structural relatedness. It is technically far easier and less laborious to perform a quantitative immunological test than to accomplish a complete sequence analysis of a protein. Indeed, for the rapid semi-quantitative assessment of the sequence relatedness of proteins there is at the moment no technique that approaches the ease and facility of immunological techniques. One method that has been extensively used in this regard is that of quantitative microcomplement fixation and one system extensively studied has been egg white lysozyme and its rabbit antibodies (350, 351). What emerges from these studies as well as from the quantitative cross-reactions of a number of other proteins of defined structure are the following observations. If one relates the amount of serum required to achieve the complement fixation reaction of the homologous protein by the cross-reacting protein (index of dissimilarity) to the percent cross-reaction in quantitative precipitin tests, the relationship proves to be linear. Thus either of these quantitative methods can be used for the study of sequence relationships among a homologous set of proteins. Moreover, if one relates the percent cross-reaction to the amino acid differences existing between the cross-reacting and homologous proteins a proportional relationship can be demonstrated. In addition, reciprocal immunization of various lysozymes in rabbits yields very good agreement from whichever direction one chooses to make the antigenic comparison. For example, turkey reacts with anti-duck A serum 62% while duck A cross reacts with anti-turkey serum 71%. Ring-necked pheasant (RNP) lysozyme cross-reacts with anti-turkey serum 86% while in the reciprocal case the cross-reaction is 83%. Indeed for seven different lysozymes studied in this way the reciprocal cross-reactions were in excellent agreement.

There are two examples in the literature where the relationship between immunological cross-reactivity and sequence relationship are not so nicely proportional. These are both small proteins with a probable smal number of antigenic sites. These are the human cytochrome c system (278 and the myoglobin system (24). With human cytochrome c it is known that there are only four antigenic sites (467) and that two of these sites tend to dominate the quantitative aspects of the reaction (368). Thus, a very small number of sequence changes can have a disproportionate effect on quantitative aspects of antigenicity. For example in the cytochrome c systems, if a single amino acid change can inactive an antigenic site (which it in fact does) as much as 25% of the reactivity is ablated. This leads to discontinuous changes in reactivity as a functio of sequence change. On the other hand, proteins with eight or more sites can experience no more than a maximal change of 10-12% in reactivity accompanying single amino acid sequence changes. The fact that lysozyme, which is a small protein, experiences no such discontinuities in the relationship of cross-reactivity and sequence relatedness may have at least two possible explanations. One is that despite the small size of lysozyme it has a relatively large number of independent antigenic sites. As the number of sites is in some measure related to the number of sequence differences between immunogen and the sequences of the same protein in the immunized specie, it would be predicted that egg white lysozyme would invoke the production of many more different antibodies than would human cytochrome in rabbits.

Human cytochrome c differs from rabbit cytochrome c by only 12 residues out of a total of 104. While the sequence of rabbit lysozyme is not known it can be estimated that of its 129 residues at least 40-50 would be different from that of the chicken lysozyme. Thus the potential for many antigenic sites and much intrasite heterogeneity exists if such phenomena depend on the sequence differences between egg white and rabbit lysozymes. It is interesting that the antibodies against a singl site in lysozyme which can be isolated by a peptide immunoabsorbent exhibit gross electrophoretic heterogenity (8 bands) while similar single site fractions isolated from anti-cytochrome c serum yield a very small number of bands (1-2) (368).

Another possible explanation for the smooth relationship between cross-reactivity and sequence differences in the lysozyme system is that fortuitously the sequence changes which occur in closely related species of lysozyme do not have such an important quantitative effect on the structure of individual determinants.

Whatever the reasons for the success of the immunological method in predicting structural relatedness among lysozymes the chances are quite good that in general with relatively large proteins (greater than molecular weight 30,000) the sequence immunological reactivity correlatic should hold. One might ask why the relationship between sequence change and immunological reactivity among related proteins does hold. The success of that relationship rests on three separate but related phenomena. First, antibodies react primarily if not exclusively with the amino acid side chains of surface residues. Secondly, protein evolution involves surface amino acid residues to a far greater extent than it involves internal residues. Finally, quantitative immunological techniques detect at least half of all the variation in amino acid residues that occur at the surface of globular proteins. The data underlying the first and last statements are discussed in reference (368) while the evolution matter is analyzed elsewhere (350, 351). Given the solidity of these propositions it follows that immunological techniques can provide a statistical measure of sequence relatedness among proteins.

In summary the present usefulness of quantitative immunochemical techniques in assessing the structural relatedness seems assured. The future application of such methods promises to explore not only a statistical sequence relationship but the ability to identify specific sequences in related proteins of a homologous family of proteins.

Polarization of Light and Protein Structure

S. B. NEEDLEMAN

I. Introduction

In the following chapter (Chap. 6) it will be seen that the configuration of a protein in space can be "visualized" by mathematically examining the reflections produced when a beam of X-ray is impinged on the lattice planes of the crystallized protein. While it is possible to assign loci to specific amino acids on the basis of this reflection data, in general, it is more likely that previous knowledge of the primary structure of the protein will be necessary for the interpretation of the X-ray diffraction patterns. Nevertheless, with increasing experience, X-ray diffraction probably will become a primary procedure for protein sequencing, requiring the application of chemical or some physical methods only to resolve particularly difficult short segments.

Another method, presently of only limited application to protein sequencing, is based on the additive rotation (99) of polarized light by certain amino acids in sequence as these interact with the solvent environment. A brief discussion of the principles of optical rotatory dispersion and circular dichroism will illustrate the potential contribution of these techniques of the resolution of protein structure. These topics have been thoroughly reviewed by Jirgensons (217), Moscowitz (308) and Blout (56). A more superficial treatment of the subject is given by Urry (444, 445). The present communication is based largely on these reviews.

II. Plane Polarized Light

When a beam of monochromatic linearly polarized light impinges upon a solution, a number of events are likely to occur which can be related to the structure of the solute in solution. Thus, one can detect reflection of the beam from the surface of the solution, refraction cause by slowing of the ray in passing through the solution, rotation of the plane of polarization, and absorption. The relationship between these phenomena and molecular structure will be discussed later. First, we must define the parameters which will be applied to that end.

A beam of ordinary light, or any propagated electromagnetic field, is characterized by an electric vector E and an associated perpendicular magnetic vector M. If the light beam is monochromatic, it will vibrate with a definite and constant frequency ν. The electric vector may be expressed by

$$E = E_o \cos \omega t$$

where E_o is the maximum amplitude of the wave, $\omega = \pi \nu$ and t is time. A similar expression exists for M, the magnetic component.

The two vectors propagate at right angles to each other and remain in phase. Passage of the beam through a nonabsorbing isotrophic medium will result only in slowing of the wave (refraction) and diminution of its electric component. A nonisotrophic medium (solution of asymmetric molecules) affects both the electric and magnetic components of the wave resulting in optical activity in addition to refraction.

III. Circularly Polarized Light

To return to the linearly polarized beam with wave length λ, one can visualize the longitudinal vector of sinusoidal oscillation to occur along the x axis with the amplitudes of vibration being plotted along the y axis. Thus, the xy plane is said to be the plane of vibration and such a wave (or beam) is plane polarized.

The electric field vector E may be considered as being the resultant of two equal subvectors, the E_R vector which rotates in the clockwise or "right" direction and the E_L vector which rotates counterclockwise (left) with frequency ν such that both subvectors proscribe equal but opposite angles along the x axis. Thus, E_R represents the electric field vector for a right circularly polarized wave of frequency ν and E_L, similarly, represents the electric field of a left circularly polarized wave. No rotation of the plane polarized wave will be observed if these two components pass with equal velocity through a solution medium. On the other hand, if the solution consists primarily of asymmetric molecules, the refractive indices of the solution for the right and left circularly polarized light at constant wavelength will be different. Under these circumstances, the waves of right and left circularly polarized ligth will be transmitted through the solution with different velocities and the resultant wave of plane polarized light will emerge with a phase shift. Thus, optical rotation will occur in favor of the direction of the faster component. The decrease in amplitude of the electric component is a function of the factor $3/(n^2 + 2)$, where n is the refractive index. The angle of rotation of the plane of polarization α per unit length is proportional to the difference in the refractive index for the left and right wave of circularly polarized light:

$$\alpha = \Pi/\Lambda \ (n_L - n_R)$$

where λ is the wavelength of the light beam, n_L and n_R represent the different indices of refraction of the medium for left and right circularly polarized light, respectively, and α is expressed in radians per centimeter. The difference in refractive indices need be only very small to produce a significant rotation (98). The angle of rotation at a constant frequency λ must be proportional to the number of asymmetric molecules in solution; α can be converted to the more commonly used dimensions according to the following:

$$\text{Specific rotation } [\alpha] = \frac{\alpha}{C_1} \cdot \frac{1800}{\Pi} \quad \text{and}$$

$$\text{Molecular rotation } [\phi] = [\alpha] \frac{M}{100} = \alpha \frac{18M}{\Pi C_1}$$

where C_1 is the concentration of optically active solute in g per cm^3 of solution and M is its molecular weight.

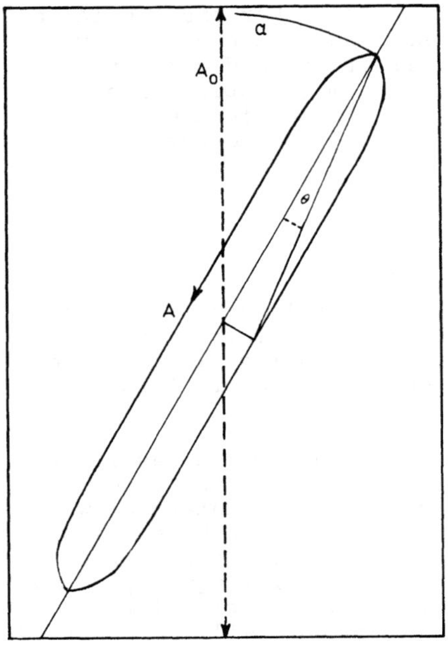

Fig. 1. Graphic definition of ellipticity θ and optical rotation of polarized light. The dashed line A_o represents the plane polarized monochromatic beam striking the sample. The transmitted beam A is elliptically polarized. Adapted from Jirgenson (47)

IV. Circular Dichroism

The rotation of plane polarized light (appearance of optical activity) as the result of differences of refractive indices for the right and left circularly polarized wave components holds only in those cases in which absorption of light by the medium is negligible. If this phenomenon is observed near an absorption band, the differences in extinc tion coefficients also must be considered. When the absorption becomes a significant factor, the medium exhibits circular dichroism and the light is said to be elliptically polarized (see Fig. 1). That is, if in addition to differences in velocity for the right and left circularly polarized beams, there is also a difference in absorption, the emergent light is elliptically polarized.

In the above discussion, it was assumed that the wavelength of the monochromatic beam of polarized light remained constant. If one were instead to measure the circular dichroism (unequal absorption) and the unequal velocity of transmission of right and left circularly polarized light over a range of wavelengths, one would note a drastic change in optical activity (rotatory dispersion) of a solute in solution with the extent of change becoming greater as one approached the absorption band of the solute. Thus, if one were to plot α, the optica activity, against wavelength λ, beginning in the region of long wavelength and proceeding towards the short wavelengths, the optical acti ity would change, first slowly, and then as one approached the UV absorbing region, more rapidly, giving maximal change in the spectral region of absorption of the substance. If one continues the rotatory dispersion measurements through this region of spectral absorption, it can be seen that the rotation reaches a peak, drops rapidly throug a corresponding through and then gradually increases again as one passes beyond the region of maximal absorption. This phenomenon is known as the "Cotton effect" (127, 302) and provides one means of characterizing certain structural aspects of a substance. The mean of wavelengths at which the peak and through occur coincides with the point of zero optical rotation and is essentially the wavelength of maximal UV absorption. This latter is referred to as an "optical active absorption band" since not all absorption bands of a given chromophore exhibit circular dichroism.

V. Drude Equation

The relationship between optical rotatory dispersion and wavelength at regions away from an optically active absorption band can be expressed by a simplified form of the Drude equation (183, 268):

$$[\alpha]\lambda = A/(\lambda^2 - \lambda_c^2)$$

in which λ_c is identical to the wavelength of any one of the transitions resulting in optical activity and A is a constant characteristic of the system. While, under ordinary circumstances, the solution of the Drude equation is difficult, a graphical mode of solution has been proposed (157). The plot of $1/[\alpha]$ against λ^2 generally results in a straight line curve from which λ_o can be calculated.

In practice, optical rotation in the region of presently accessible instrumental measurment (700-250 nm) is determined by means of a recording spectropolarimeter and the simplified (single term) Drude equation solved graphically for λ_o, an optically active absorption band.

VI. Optical Properties of Amino Acids

Over the past two decades, proteins and polypeptides have been described in terms of three major conformational states: the α-helix, β-pleated sheet and random coil. In general, the atomic and chromophoric interactions in proteins are sufficiently well-defined to allow assessment of circular dichroism measurements in terms of likely conformational content of the protein (82, 150, 188, 230, 379).

In the case of amino acids, the absorption bands producing optical rotation are too intense to permit convenient measurement or these are located in the region below 230 nm, seldom attainable with customarily available instrumentation. Nevertheless, much has been learned about the intrinsic rotatory properties of the amino acids which has been of value in interpreting the rotatory properties of the constituent polypeptide.

The optical activity of proteins is the result of asymmetry of the polypeptide chain and the asymmetry of the α-carbon atom in the component amino acids. Optical rotation of the amino acids is a function of a number of parameters including pH, substitution, the wavelength of measurement and the absolute configuration of the amino acid.

Acidification of aqueous solutions of several amino acids results in a similar shift of rotation, the direction of which is related to the absolute configuration. This was investigated systematically by Lutz and Jirgensons (270, 271) who found that increasing the concentration

Table 1. Molecular rotation, $[M]_D$, of natural L-α-amino acids in aqueous solutions and in 5N HCl

Amino acid	Molecular weight	$[M]_D$ in H_2O	$[M]_D$ in 5N HCl
Alanine	89.1	+1.6	+13.0
Valine	117.1	+6.6	+33.1
Leucine	131.1	-14.4	+21.0
Isoleucine	131.1	+16.3	+51.8
Serine	105.1	-7.9	+15.9
Threonine	119.1	-33.9	-17.9
Cysteine	121.1	-20.0	+7.9
Methionine	149.2	-14.9	+34.6
Aspartic acid	133.1	+6.7	+33.8
Glutamic acid	147.1	+17.7	+46.8
Lysine	146.2	+19.7	+37.9
Arginine	174.2	+21.8	+48.1
Histidine	155.2	-59.8	+18.3
Tyrosine[a]	181.2	-59.8	-18.1
Tryptophan	204.1	-68.8	+5.7 (1N HCl)
Proline	115.1	-99.2	-69.5

[a]Essentially insoluble in water; the optical activity in 5N HCl is more positive than in dilute HCl. Adapted from Greenstein and Winitz (166).

of an acid produces a more positive optical rotation of an L-antipode. This is related to the degree of suppression of dissociation of the carboxyl group. Certain amino acids (e.g., histidine, leucine and tryptophan) are levorotatory in an aqueous solution but dextrotatory in acidic solutions. The change in rotation on acidification is the same in all cases. Greenstein and Winitz (166) expressed the shift rule according to:

$$\text{L-amino acid} \quad [M]H^{+} - [M]_{H_2O} = + \ \Delta [M]$$
$$\text{D-amino acid} \quad [M]H^{+} - [M]_{H_2O} = - \ \Delta [M]$$

The molar rotation of the common amino acids in water and in 5N HCl is given in Table 1.

Spectral analysis of alkanes into the UV show that absorption is negligible above 140 nm and diffuse below this wavelength. As a result, the major contribution to rotation of some amino acids arises from the carboxyl group (54).

L-Amino acids with no chromophore or with only a weak one at the β-carbon atom give simple optical rotatory dispersion curves which ascend positively with decreasing wavelength (391). Amino acids with strong chromophores at the β-carbon show anomalies in the visible and near UV regions.

Most amino acids have substituents that absorb between 160 and 300 nm which contribute to the optical rotation as a function of their distance from the center of asymmetry. Those amino acids with chromophoric substituents beyond the β-carbon give plain curves similar to those for aliphatic derivatives of alanine. In general, all of the common L-amino acids show a positive Cotton effect near 210 nm with a peak at 225 nm and a through at 193 nm when measured at pH 1 (208). Cystine, more optically active because of the asymmetry of the disulfide bond, exhibits a deep trough at 206 nm and the entire curve lies in the negative part of the chart. Tyrosine and tryptophan have additional Cotton effect near 275 nm.

VII. Optical Properties of Peptides

There is little correlation between the optical rotation of a free amino acid and its residual rotation in a protein chain (54). Nevertheless, when a free amino acid is incorporated into proteins, amino acids similar enough to give similar dispersion curves should undergo shifts of the same sign and order of magnitude (99). For peptides or proteins containing such amino acids, the rotatory properties will depend more on the backbone conformation than on the amino acid composition. Conversely, rotatory properties contributed by chromophoric substituent groups in the form of free amino acids will retain these properties after incorporation of the amino acid into the peptide chain. Thus, it is possible, to some degree, to gain insight as to the amino acid content of segments of peptides based on their optical rotatory measurements.

Optical rotation varies as a function of chain length in peptides containing two to six amino acid residues; the effect of end groups or "chain ends" is negligible in the large polypeptides or proteins (376). Furthermore, polypeptides frequently contain helical, β-extended and random conformations within the same molecule and each structural entity will contribute independently to the variation in rotatory properties of the molecule with conformation change (99). For a totally random or denatured structure, no effect is seen in the optical rotatory properties when the chain length is increased beyond about 20 amino acid residues. On the other hand, it has been shown (55, 385) that helical structures may be formed in organic solvents with a few as six or seven appropriate amino acid residues (36) and interpretation of optical rotation information is quite dependent upon chain length in small peptides (379) or in large globular proteins which may contain short helical segments stabilized by disulfide or ionic cross-linking (218). The β-conformation of proteins represents an unusual situation in that, unlike helical structures, they possess no "screw sense" yet show significant effects on the rotatory properties of the protein in which they appear.

It is apparent that as one modifies the solvent environment of a protein from one favoring helical formation to one promoting randomization, a transition in optical rotation is to be expected as the helical or β-pleated sheet segments are replaced by random ones (379). Changes in temperature away from some optimum value contribute to the randomization process and, hence, to variation in the optical rotatory dispersion. In addition, one might expect some slight degree of influence of molecular weight on the rotation of polarized light, but only as the molecular weight of a protein affects the stability of a particular conformation (390).

VIII. Conclusion

Of greatest interest in the context of the present chapter is the role of composition, that is, the number and kind of amino acid residues present in the protein as well as the sequence, at least of the terminal residues, as these relate to optical rotatory dispersion. It was stated at the beginning of this chapter that ORD and CD would not constitute a primary technique for resolving the sequence of a peptide segment. While the reasons for this assertion should now be evident, it is nevertheless now possible to apply the techniques discussed here to gaining some insight into the order of amino acid residues, at least at the terminal regions of peptides. Sufficient information is now in hand to permit the assignment of one of many amino acids to terminal or penultimate positions with a high degree of confidence (217). With experience perhaps one or two additional positions can be labeled, though with less enthusiasm than the terminal sites.

The X-ray Crystallography Technique in Protein Sequencing

K. K. KANNAN

I. Introduction

Since the first X-ray diffraction photographs of protein crystals by Bernal, Crowfoot and Hodgkin (39), X-ray crystallographic techniques for studying complex protein structures have been tremendously improved. This is amply illustrated by the detailed analysis of the structure of myoglobin by Kendrew and co-workers (231).

The high resolution structure of sperm whale myoglobin was elucidated by X-ray diffraction methods before the amino acid sequence was fully established. In the classical paper by Kendrew et al. (232), the structure-sequence relationship was well discussed. They also laid down a number of criteria for identifying the amino acid sequence from X-ray diffraction analysis of protein crystals. Since then, more than 50 protein structures have been solved to high resolution by X-ray crystallographers (Table 1) and the validity of these principles was vindicated in all instances.

A number of questions arise about the possibilities of X-ray diffraction methods in sequence analysis. How much knowledge of the amino acid sequence of the protein does a crystallographer need in order to interpret his results, or can the interpretation be carried through without recourse to the primary structure? Has partial amino acid sequence information any value in establishing the complete sequence from X-ray diffraction work? The experience of the protein crystallographers during the past 16 years on high resolution protein structure analysis is unambiguous on these points. The successful detailed interpretation of high resolution protein structures, though possible to a limited extent without the amino acid sequence, is indeed extremely dependent on the availability of at least the amino acid sequences of a complete set of peptides determined by chemical methods. Cooperation between the protein crystallographer and the sequence analyst can speed up the elucidation of the complete amino acid sequence.

The intention of this article is to explain the protein crystallographer's contribution in establishing the amino acid sequence by X-ray diffraction methods and to analyze the correlation between structure and sequence for some proteins.

II. Outline of the Protein Crystallographic Method

As a number of articles (137, 284, 343) have given detailed discussions of the methods of X-ray diffraction and, in particular, their application to protein structure analysis, the purpose of this section will be only to introduce the essential concepts. The interested reader is directed to the literature for further study.

Table 1

No.	Protein	Species	Resolution	Year	Ref. structure	Ref. sequence	Comments on structure-species correlation
1	Adenylate kinase	Porcine muscle	3.0	1974	397	182	
2	Alcohol dehydrogenase	Horse liver	2.5	1974	138	220	
3	Bence-Jones protein	Human	3.5, 2.2	1974	393, 142	436	
4	Carbonic anhydrase B	Human	2.2	1975	226	17, 262	The sequence determined by Nyman and co-workers (17) was compatible with the structure. The sequence first reported by Lin and Deutsch (262) was incompatible with the structure.
5	Carbonic anhydrase C	Human	2.0	1971	224, 261	184	Only fragmentary sequence information was available when the electron density map was interpreted. The completed amino acid sequence was compatible with the structure.
6	Carboxypeptidase A	Bovine	2.0	1967	359	62	Only partial amino acid sequence was available when the structure was established. An attempt to determine the sequence from the electron density map of the unknown regions of the chemica-sequence was correct to 60%.
7	α-chymotrypsin γ-chymotrypsin chymotrypsinogen	Bovine	2.0, 2.8	1967 1971 1970	286, 441 398 156	178	
8	Concanavalin A	Jackbean	2.0	1972	131, 177	131, 102	Sequence of cyanogen bromide fragments were aligned on the basis of X-ray diffraction analysis.
9	Cytochrome b_5	Calf liver	2.0	1971	287	334, 439	The sequence of a dipeptide Ile 13 - Glu 14 proposed by Tsugita et al. (439) was confirmed from the electron density map.
10	Cytochrome c	Horse	2.8	1971	115	277	
11		Bonito	2.8	1971	115	314	
12		Tuna	2.4	1971	426	427	

13	Cytochrome c_2	R. rubrum	2.0	1973	387	128	
14	Cytochrome c_{550}	M. denitrificans	2.5	1975	435	460	Sequence of tryptic peptides arranged on the basis of the structure and verified by comparing the amino acid composition of chymotryptic peptides.
15	Elastase	Porcine	3.5	1970	460	410	
16	Fab'	Human	2.8	1973	347	242	Tentative sequence derived for the V_H chain on the basis of homology to other Bence-Jones proteins and electron density maps.
17	Ferrodoxin	P. aerogenosa	2.0	1973	6	428, 440	
18	Flavodoxin	Clostridium MP	3.25	1972	14	-	Only amino acid composition known when structure was determined.
19	Flavodoxin	D. vulgaris	2.5	1972	458	-	
20	Glyceraldehyde-3-phosphate dehydrogenase	Lobster	3.0	1974	71	105	
21	Hemoglobin	Bloodworm	2.5	1971	267	210	
22	(Erythrocrueorin)	Chironomus	2.5	1971	341	67, 414	
23		Horse	2.8	1968	205	74	
24		Lamprey	2.0	1971	267	259	
25	High potential iron protein	Chromatium	2.0	1971	85	129	
26	Insulin	Porcine	2.8	1969	3	381	
27	Lactate dehydrogenase	Dogfish	2.8	1970	4	430	Only fragmentary sequence information available when the structure was determined. Available peptide sequences were located on the basis of the electron density map.
28	Lysozyme	Hen eggwhite	2.0	1965	47	81, 219	The sequence according to Canfield (81) was found compatible with the X-ray determination.
29		T_4 Bacteriophage	2.5	1974	290	438	

Table 1 (continued)

No.	Protein	Species	Resolution	Year	Ref. structure	Ref. sequence	Comments on structure-sequence correlation.
30	Malate dehydrogenase	Porcine	3.0	1972	190		Only fragmentary sequence was known when the structure was interpreted.
31	Myogen	Carp muscle	1.85	1972	329	329, 340	Sequenced tryptic peptides aligned on the basis of the electron density map and sequence homology to hake myogen (340).
32	Myoglobin	Sperm whale	2.0	1960	231	136	Only partial sequence was known when the structure was determined.
33	Nuclease	S. aureus	2.0	1971	22	429	
34	Pancreatic trypsin inhibitor	Bovine	2.5	1970	204	228	
35	Papain	Papaya latex	2.8	1968	125	301	Structure interpretation revealed errors in the first amino acid sequence analysis, which led to a redetermination of the sequence.
36	Prealbumin	Human	2.5	1975	49		
37	Phosphoglycerate kinase	Horse	3.0	1974	48		
38	Phosphoglycerate mutase	Yeast	3.0	1974	76		
39	Rhodanase	Bovine liver	3.9	1973	412, 413		
40	Ribonuclease A Ribonuclease S	Bovine	2.0	1967 1970	227, 474	416	
41	Rubredoxin	C. pasteurianum	2.0	1970	186	292	
42	Subtilisin BPN'	B. amyloliquefaciens	2.5	1969	472	279	
43	Subtilisin novo	B. subtilis	2.8	1971	126	332	
44	Superoxide dismutase	Bovine	3.0	1975	374	418	
45	Thermolysin	B. Termoproteolyticus	2.3	1972	288	437	

46	Thioredoxin	Escherichia coli B	2.5	1975	195	194
47	Triose Phosphate isomerase	Chicken muscle	2.5	1975	30	30, 161
48	Trypsin	Bovine	2.7	1974	421	299
49	Trypsin-soyabean trypsin inhibitor complex	Porcine-soyabean	2.5	1974	57	237, 299
	also PTI complex	Bovine	2.8 (1.9)	1973	378	228, 299

Notes to Table 1:

1. Comments on structure-sequence correlation is made only when the structure interpretation from the electron density map has

a) provided an independent X-ray sequence
b) been used to place sequenced peptides
c) been useful in identifying functional residues

2. Comments are not made if the sequence correlated well with the structure or if no correlation between structure and sequence are available.

The first stage in the crystal structure analysis of a protein is, of course, to produce suitable single crystals of the protein in which one is interested. As has been aptly pointed out by Eisenberg (137), this stage of protein crystallography is still an art rather than a science. Once the protein crystals of the required dimensions and quality have been obtained, the protein crystallographer's attention turns to the production of heavy atom derivatives so as to utilize the method of multiple isomorphous replacement to solve the "phase problem" (137, 284 343). This method requires that the cell dimensions and space group are the same for the native protein and the heavy atom complexed crystals. The locations of the heavy atoms bound to the protein in the unit cell are then determined and utilized to calculate the phase angles. The electron density map is then computed for the native protein, and contains information about the atoms that build up the protein. The electron density map thus provides certain information about the amino acid sequence of the protein.

III. Some Important Concepts in X-ray Diffraction Analysis of Protein Crystals

A protein crystal is a three-dimensional network of a large number of protein molecules. The unit cell may contain one or more protein molecules. Normally about 40-60% of the unit cell volume is occupied by unordered solvent molecules (284, 285), which gives the protein molecules in crystals more or less the same aqueous surrounding as they have in solution. X-ray diffraction from a protein crystal can be considered to be a three-dimensional network of reflections with integral coordinates h, k and l and a definite intensity I(h,k,l) that can be recorded physically. The amplitude $|F(h,k,l)|$ is related to the intensity by the expression:

$$F^2(h,k,l) \propto I(h,k,l) \tag{1}$$

However, the crystallographer is interested in the structure factor F(h,k,l), which is a complex quantity, and can be written as

$$F(h,k,l) = |F(h,k,l)| \exp\left[i\,\alpha(h,k,l)\right] \tag{2}$$

where $i = \sqrt{-1}$.

The value α (h,k,l) is called the phase angle of the reflection and is not directly measurable, but can be determined by the method of multiple isomorphous replacement with a set of heavy atom derivatives. The electron density of the protein at any point (x,y,z) in the unit cell is given by the expression:

$$\rho(x,y,z) = \frac{1}{V} \sum_h \sum_k \sum_l |F(h,k,l)| \exp\text{-}i\left[2\pi\,(hx + ky + lz) - \alpha(h,k,l)\right] \tag{3}$$

where V is the volume of the unit cell. Once the phase angles α (h,k,l) are known for the reflections, the electron density map can be easily calculated. The "resolution" of the electron density map gives an estimate of the details one can identify in these maps. An electron density map at 2 Å resolution gives more detailed information than one at 3.5 Å. The number of reflections required for a 2 Å electron density map is much greater (ca. 5.4 times) than that needed for a 3.5 Å map. The resolution obtainable from a protein crystal is dependent on the quality of the crystals used in the X-ray diffraction work.

2.8 Å - 2 Å electron density maps are usually referred to as high resolution maps and can be interpreted in terms of the atomic positions. It is this stage of the analysis which will be considered here in some detail in relation to the amino acid sequence of the protein.

IV. Building up the Electron Density Map

The triple summation [Eq. (3)] for each point x, y, z in the unit cell involves many thousands of reflections for a high resolution electron density map of a protein molecule, and is computed using a digital computer. Usually the electron density values are calculated as a number of two-dimensional planes with one of the coordinates x, y or z kept constant and the remaining two varied one at a time. The distance between two successive points is usually about one-third of the resolution or less. These planes of electron density are then contoured manually or with a computer by connecting points with equal electron density and the contours are transferred to transparent plastic sheets. The lowest level chosen for contouring and the interval for contouring are very important for the eventual interpretation of the electron density maps. If the lowest level, which is usually equal to about twice the increment between the contour levels, is chosen too low, then the electron density maps will include too much detail and will become very messy. If this level is chosen too high, then the maps will be discontinuous and consequently difficult to interpret. Usually about 0.45 electrons/A^3 above the mean electron density seems to give a balanced lowest level contour. The two-dimensional electron density maps thus produced are then stacked together with the correct separation defined by the third dimension. A typical three-dimensional electron density map might consist of about 50 sections of two-dimensional planes for a protein with a molecular weight of about 30,000. These stacks of electron density maps are the basis for interpretation in terms of the molecular structure of the protein. To make the process of interpretation easy, usually 7 to 10 sections of the electron density map are viewed at a time. In the following discussion all the illustrations are mono- or stereo-photographs of such composite maps.

V. Interpretation of the Electron Density Map

A. Molecular Boundary

As mentioned earlier, the protein crystals consist of the ordered protein molecules as well as a large amount of more or less disordered solvent molecules. To begin the interpretation of an electron density map, the molecular boundaries must be located. Figure 1 shows such a molecular boundary from the high resolution map of human carbonic anhydrase C (224, 261). The ordered protein molecules have heavy continuous electron density compared to the discontinuous and low electron density of the solvent molecules, providing for a good contrast between the protein molecules and the solvent region of the crystals and thereby facilitating the localization of the molecular boundaries.

B. Main Chain

Once the molecular boundaries have been identified, the interpretation of the protein part of the elctron density map can be achieved. The

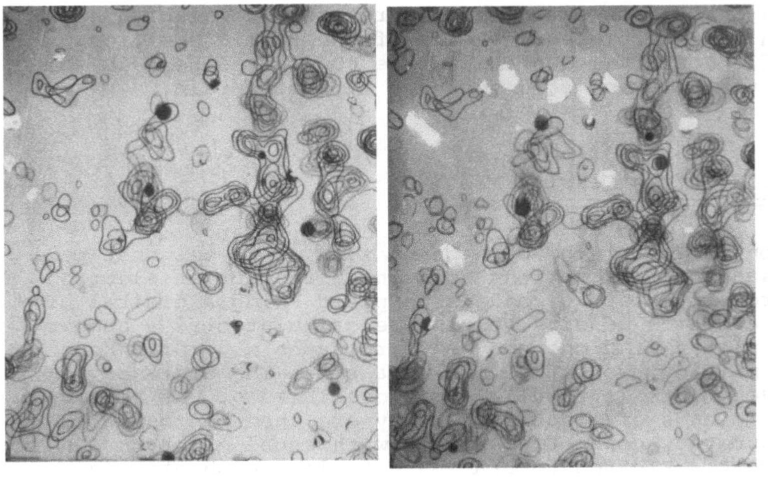

Fig. 1. Molecular boundary in human carbonic anhydrase C. The continuous heavy electron density regions represent the protein molecule and the discontinuous low electron density that of the solvent in between the protein molecules

protein polymer can be considered to be made of a peptide chain to which the amino acid side chains are attached. This concept is helpful in interpreting the electron density maps of proteins. This ribbon of the polypeptide chain in the protein is usually observed as stretches of continuous electron density generally referred to as the main chain electron density (Fig. 2). The helical protions of the protein (Fig. 3), when present are usually seen as more or less continuous spirals of electron density with side chain densities hanging out from the helix (Fig. 3). In the case of extended stretches of polypeptide folding as in a strand of β structure the electron density is seen as a continuous ribbon of electron density (Fig. 2). It is not unusual that there is a more or less continuous bridge of electron density between the carbonyl group and the amide nitrogen of neighboring strands involved in β-structure. The side chain densities along a strand in a β-

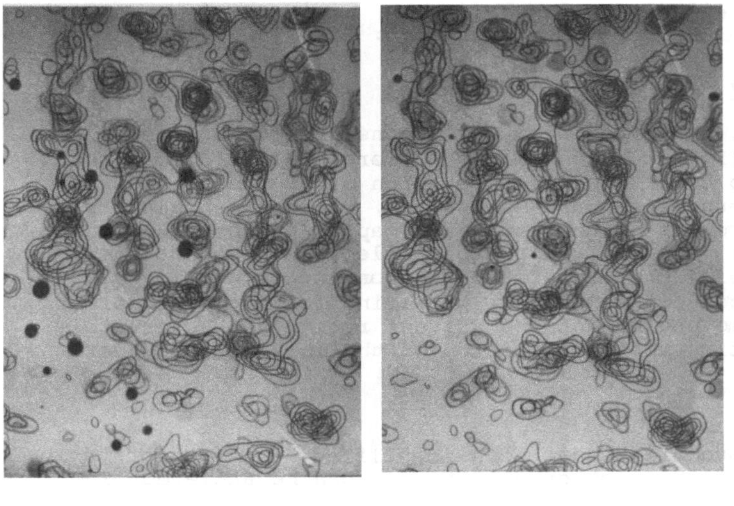

Fig. 2. Main chain electron density of a few polypeptides of human carbonic anhydrase C which form part of the central pleated sheet structure of the molecule

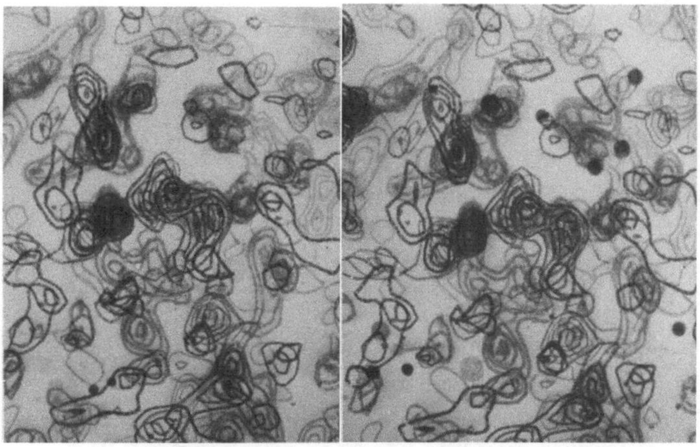

Fig. 3. A helical region of the human carbonic anhydrase B molecule

Fig. 4. Main chain electron density showing carbonyl knobs indicated by arrows

sheet (Fig. 2) will protrude alternately on either side of the sheet. The carbonyl group in the peptide bond with the electron-dense oxygen atom connected to the carbon at a distance of 1.24 Å is usually seen as a dense short knob (Fig. 4) in the main chain electron density. The characteristic carbonyl densities together with the densities for the side chains defines unambiguously the direction of the polypeptide backbone from the amino terminus to the carboxy terminus.

Even in the best electron density maps there are occasional breaks in the ribbon of electron density which complicate the interpretation of the course of the main chain folding. These broken stretches of electron density are usually interpreted separately and subsequently connected up. The connectivities of the main chain folding might sometimes be ambiguous due either to badly resolved electron density or to total lack of electron density for a few residues. The latter situation is encounterd occasionally on the surface of protein molecules and probably reflects a certain amount of flexibility in the portions of the molecule that are in contact with the solvent. The known parts of the amino acid sequence are of valuable help in resolving such ambiguities.

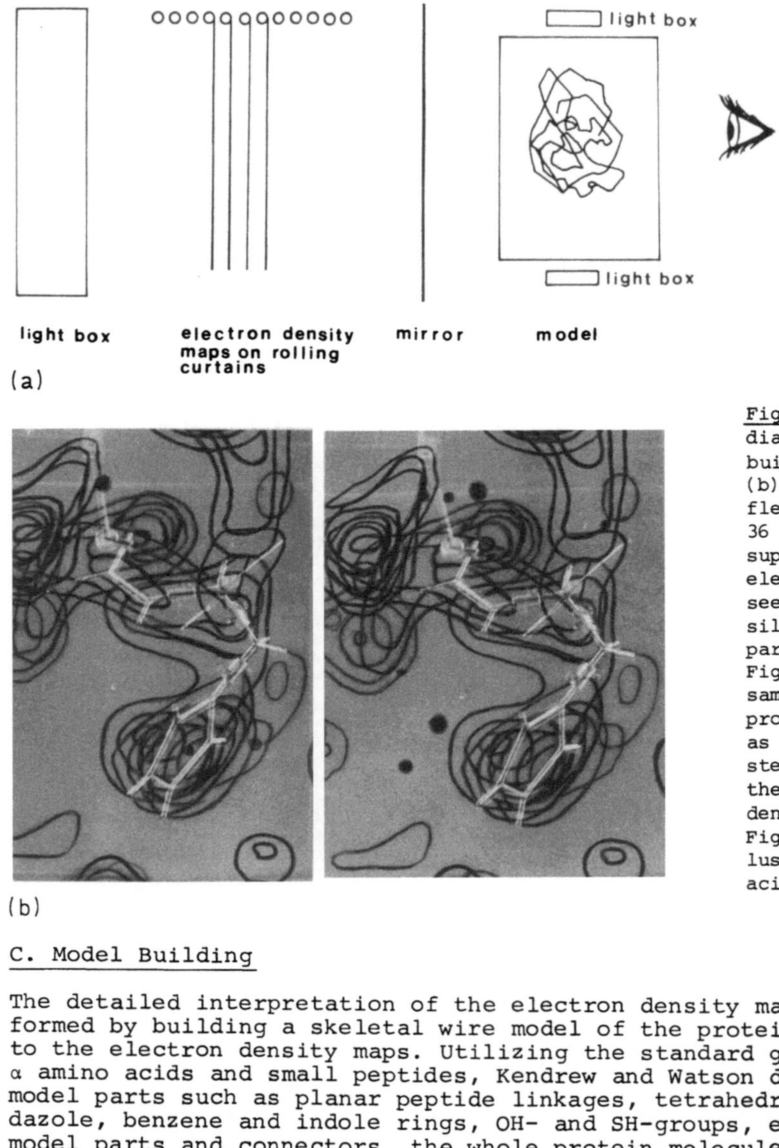

Fig. 5. (a) Schematic diagram of the model building device. (b) Stereopair of reflected image of His 36 (i.e. d-amino acid) superposed on the electron density as seen through the half silvered mirror. Compare this picture with Figure 22, where the same stereopair is reproduced with His 36 as l-amino acid. The stereo pairs of all the model-electron density superpositions Figures 6-27 are illustrated as l-amino acids.

C. Model Building

The detailed interpretation of the electron density map is usually performed by building a skeletal wire model of the protein corresponding to the electron density maps. Utilizing the standard geometries of the α amino acids and small peptides, Kendrew and Watson designed skeletal model parts such as planar peptide linkages, tetrahedral carbons, imidazole, benzene and indole rings, OH- and SH-groups, etc. Using these model parts and connectors, the whole protein molecule can be built and clamped to thin metal rods in a rigid metal frame. (Such model parts can be purchased from Cambridge Repetition Engineers Ltd., Cambridge, England).

Richards has described (373) a simple and convenient mirror system for interpreting the electron density maps and building the model simultaneously. In this device the rigid metal frame for building the model is placed on one side of a half-silvered mirror. The electron density maps, drawn to the same scale as the model parts (1 Å = 2 cm), on trans parent sheets are stretched on the other side of the mirror. A light box illuminates a stack of map sections while a thin strip of light illuminates the corresponding part of the model. Since the mirror image of the model will be viewed during the model building, the electron density sections are stacked as mirror images of the real orientation. The mirror may be aligned either 45° (373) to the map sections or parallel to the electron density sheets (97). The latter method (Fig. 5a) has the advantage that the model is accessible from all sides to a single person as he views through the mirror, which makes the model building rather convenient. On the other hand positioning the mirror at 45° to the map sections allows easy stereophotographing of the model superposed onto the maps through the mirror.

A stack of 7 to 10 sections of the electron density map is viewed through the mirror and the positions of the model parts are adjusted for the best superposition of their reflected image on the electron density map. The model parts are easily manipulated by rotating about the single bonds which are joined together by the connectors.

To begin the model-building process, a well-defined region of the electron density map, such as a helical region or a stretch of extended chain involving pleated sheet formation, is usually chosen. The main chain is fitted to the electron density and the side chains can be indicated with β carbons or as complete groups attached to the α carbon atoms and checked for good fit. This procedure is continued until a difficult place in the electron density map is reached. New regions of the maps are then explored and the model fitted to the electron density. The unconnected peptides are then tied together if there is no ambiguity involved.

D. The Side Chains

The twenty different types of amino acid residue which occur in a protein molecule are characterized by the side chain part of the residue (Figs. 6-27) and the electron densities of the side chains reflect these characteristic features there by aiding the identification of the residue.

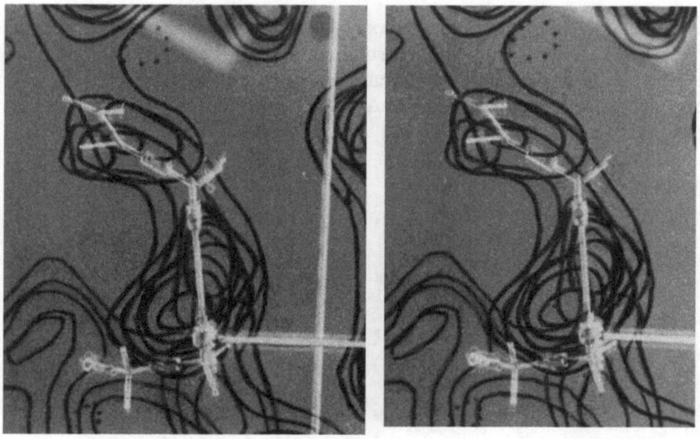

Fig. 6. Gly 98. Glycine is characterized by the lack of a side chain. The absence of any electron density in the side chain position of the α carbon thus often indicates a glycyl residue. The lack of the β carbon occasionally gives a low or a discontinuous main chain density. It is also a preferred residue in certain bend regions of the polypeptide chain where bulkier side-chains are not readily accommodated (100, 450)

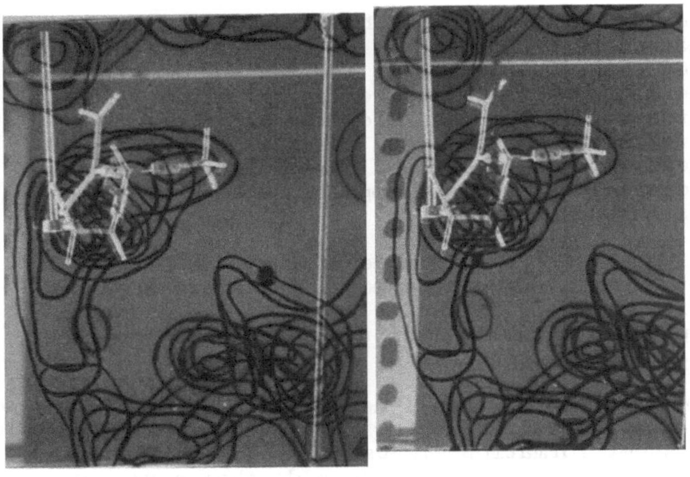

Fig. 7. Ala 38. The side chain of alanine is only a methyl group and the electron density is frequently unresolved from that of the α carbon but appears only as a protrusion from the main chain density

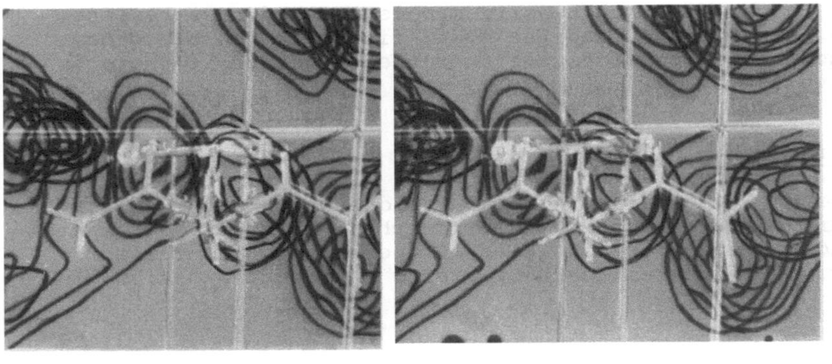

Fig. 8. Ser 197. The γ-hydroxyl group of serine is often involved in hydrogen-bonding which stabilizes the orientation of the side chain. A short electron density at the side chain position with occasional heavier density or a proturberation from the β carbon density indicating the hydroxyl group and the hydrogen bonding capability can be useful in identifying this residue

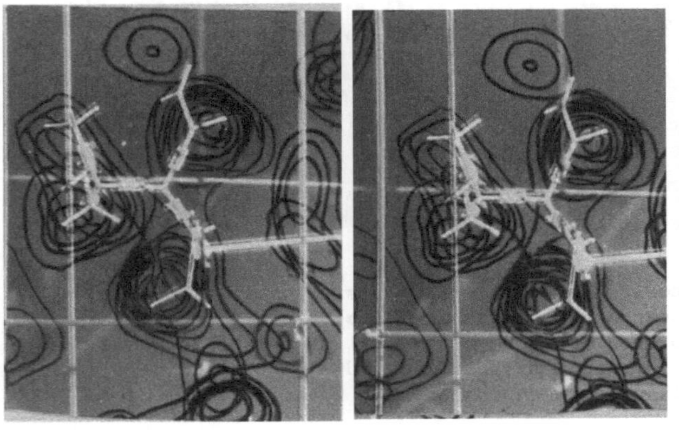

Fig. 9. Val 143. This hydrophobic residue can be identified by the characteristic forking at the β carbon atom. The hydrophobic nature of valine is sometimes helpful in distinguishing this residue from threonine

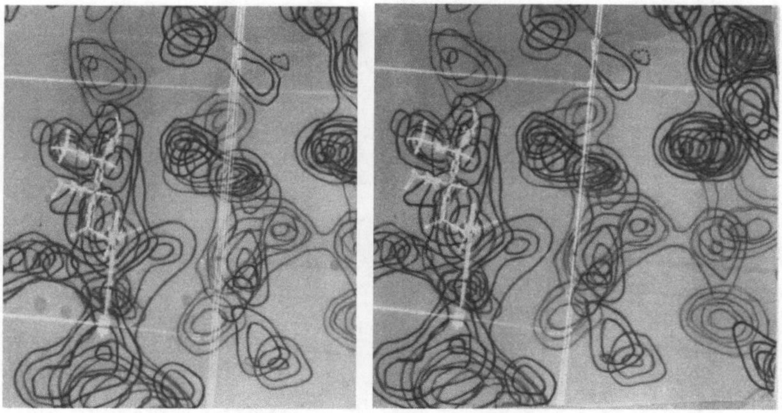

Fig. 10. Thr 193. Theronine is also forked at the β carbon. The -OH group in threonine unlike the methyl group in valine, can take part in hydrogen bonds, which property can be utilized to distinguish threonine from valine. The asymmetric β carbon atom of threonine provides a useful check when fitting this residue in the electron density maps

The interior of the molecule is usually compact and the electron density is also better defined in this region. On the surface of the molecule some side chains which are in contact with the solvent can take different statistical positions. The electron densities for such side chains are usually poorly defined and are thus often difficult to identify with a high degree of certainity.

The ring-structured residues: Trp (Fig. 26), Phe (Fig. 24), Tyr (Fig. 25), and His (Figs. 22, 23) are easily recognized by the broad electron density and are generally distinguished by the shape of the density. Among the aromatic residues Trp is often the easiest to identify from the electron density maps. The residues Val (Fig. 9), Thr (Fig. 10), Leu (Figs. 15, 16), Asp (Fig. 11), Asn (Fig. 12), Glu (Fig. 13) and Gln (Fig. 14) all have a typical forked structure and are generally recognized by their forked electron density. Ambiguities can arise between Val and Thr between Asn, Asp and Leu, and between Gln and Glu residues owing to very similar electron densities. Ile (Fig. 17) can be confused with any of the forked residues depending on the available electron density. Some of the ambiguities stated above can be resolved to a certain extent by model building and also by the fact that side chains, such as of Val, Leu, Ile, are hydrophobic residues which cannot take part in hydrogen bond formation, while Thr, Asn, Gln can participate in hydrogen bond formation, and Asp and Glu can form ion pairs with Arg, Lys, or His residues. As a general rule, charged residues that are located in the interior of the molecules form ion pairs. Of the long side chain types, Arg (Fig. 21) is more easily identified than is Lys (Fig. 20). Arg also exhibits unique electron density with a forked guanidinium group. Lys has usually poorly defined electron density, probably because it is often located on the surface of the protein molecule in contact with the solvent and has a certain amount of flexibility in its orientation. However, when Lys is involved in ion pair formation it is easily recognizable from the electron density maps. The sulfur-containing residues, Cys (Fig. 18) (Cysteine) and Met (Fig. 19), generally show high electron density for the electron-

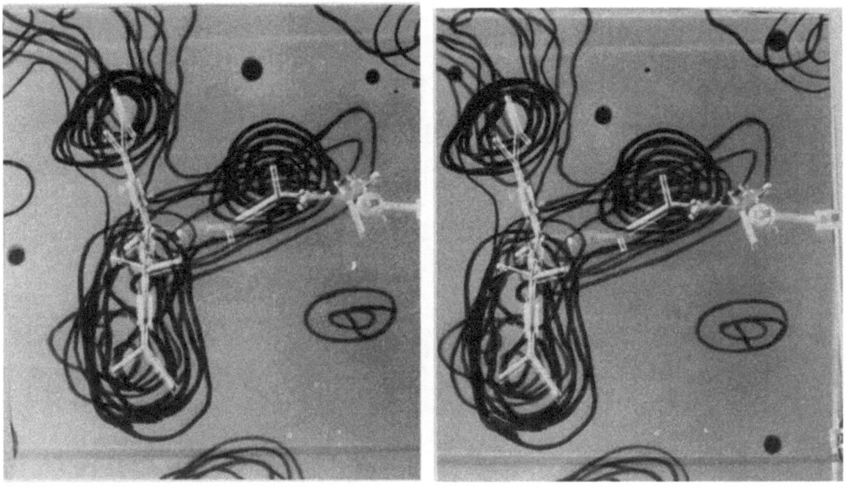

Fig. 11

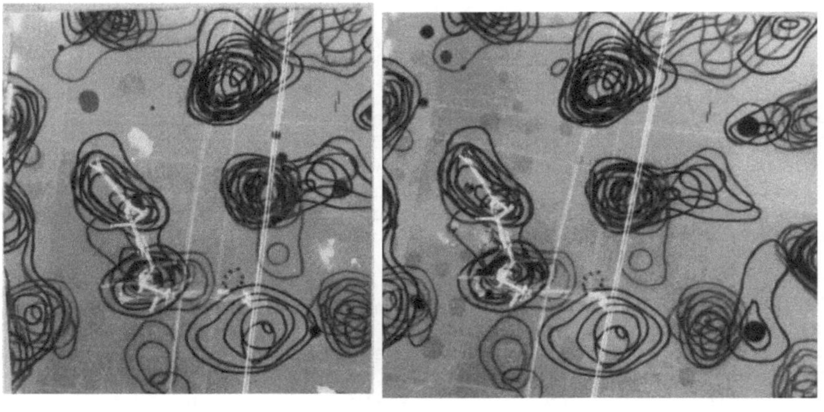

Fig. 12

Figs. 11 and 12. Asp 243 and Asn 61. Aspartic acid and asparagine are generally difficult to distinguish from the electron density alone. These residues have a planar fork at the γ carbon atom. Asp and Asn can sometimes be distinguished due to their different capacities as hydrogen bond donor and acceptor

dense sulfur atom and are recognized from this criterion and the chain length. Ala (Fig. 7) is a short amino acid having only a β carbon atom and often appears as a protrusion from the main chain. Ser (Fig. 8) is recognizable by the extra oxygen atom with its hydrogen-bonding capability. Pro (Fig. 27) is typified by its occurrence as a broad flat main chain density. This characteristic of Pro has been often useful in correctly identifying it from the electron density maps. Gly (Fig. 6) is characterized by the lack of the side chain electron density.

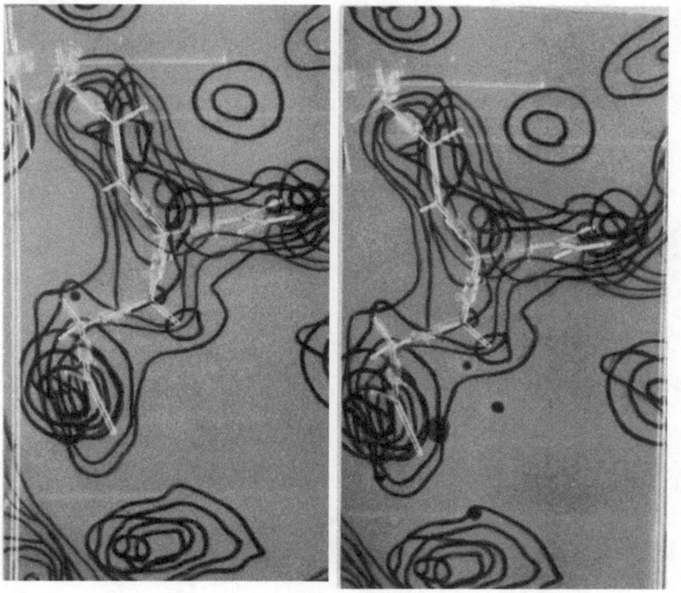

Fig. 13

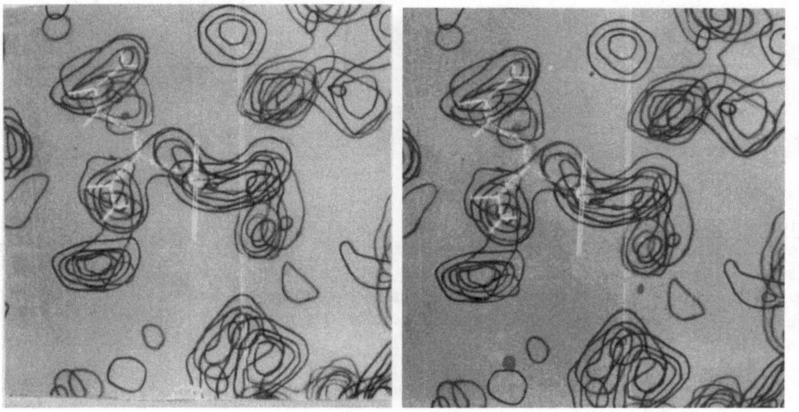

Fig. 14

Figs. 13 and 14. Glu 106 and Gln 158. Glutamine and glutamic acid side chains are also difficult to distinguish from the electron density maps alone but the hydrogen-bonding pattern can provide some help in distinguishing them as in the case of Asp and Asn

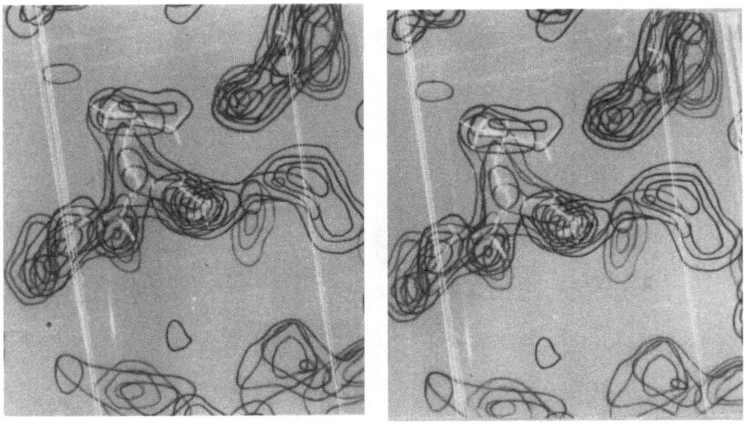

Fig. 15

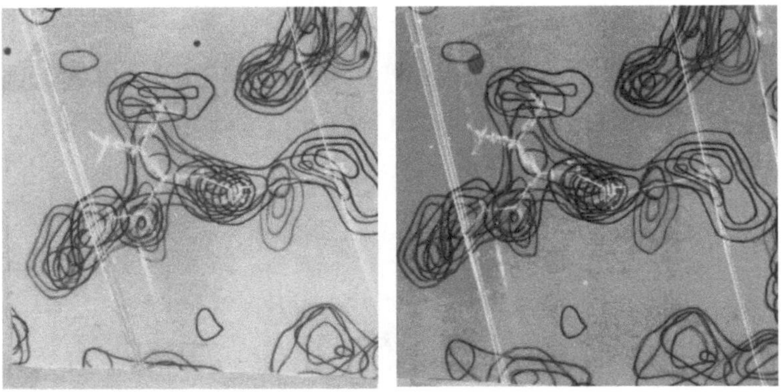

Fig. 16

Figs. 15 and 16. Leu 185. The side chain of leucine is hydrophobic and forks at the tetrahedral C γ atom. In the interior of the molecule where the hydrophobic residues usually pack very well, leucine is generally easy to identify from the electron density maps. The distance of the forked density from the main chain often distinguishes leucine from valine, as seen in Figures 15 and 16 respectively. Indeed, it was difficult to build Asp or Asn in this well-defined Leu density

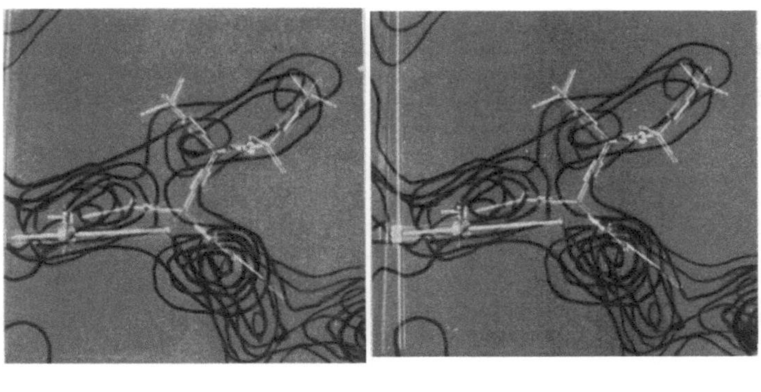

Fig. 17. Ile 216. This hydrophobic residue with one short and one long arm at the fork often shows these features resolved in the electron density maps. The asymmetry of the β carbon in Isoleucine provides a useful check when building the model in the electron density maps

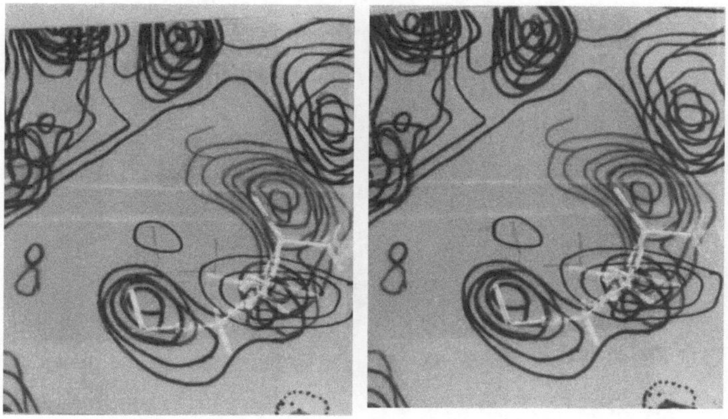

Fig. 18. Cys 206. Cysteine could be one of the easily identifiable amino acids from the electron density maps due to the electron dense sulfur atom. In favorable circumstances this residue can be located by its affinity to mercurials. The position of the mercury atom often locates the cysteine residue in the electron density maps (434).
In the case of disulfide bridges, the sulfur atoms indeed show high electron densities and are easily located. Such was the case in carboxypeptidase (264), where the presence of a disulfide bridge was first recognized on the basis of the electron density map

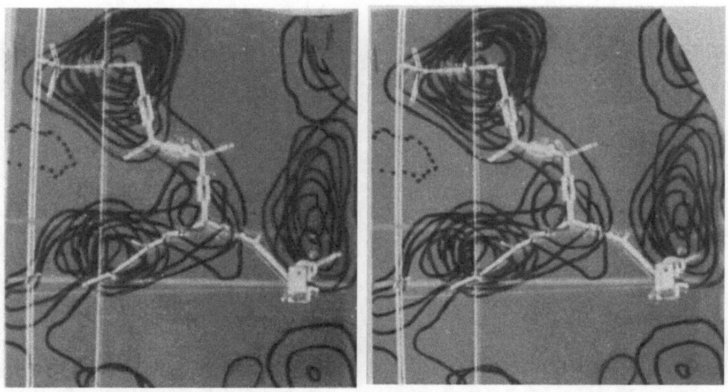

Fig. 19. Met 241. Methionine is characterized by the electron dense sulfur atom. The methionyl side chain is long with the sulfur as the penultimate atom and is relatively easy to identify in the electron density maps

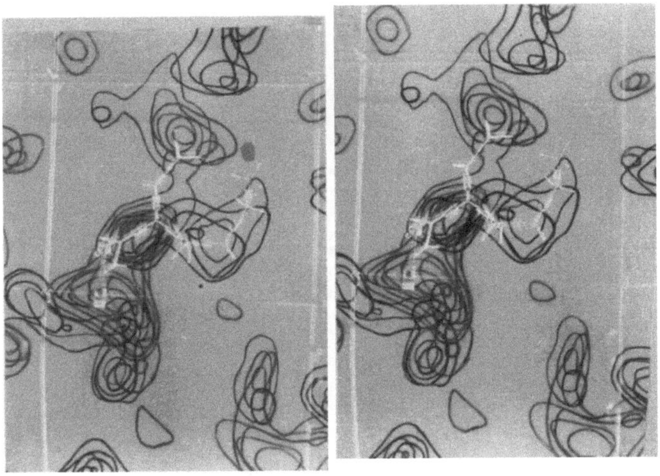

Fig. 20. Lys 154. Lysine is a long, positively charged side chain most often located on the surface of protein molecules in contact with the solvent where its flexibility can be expressed. This has made lysine relatively difficult to identify from electron density maps. However, when the side chain is more or less rigidly fixed by ion pair formation with oppositely charged side chain the interpretability of lysine is significantly improved. The δ- and ε-groups of this residue often show a statistical distribution among two or more positions, especially in forming ion pairs with neighboring charged residues

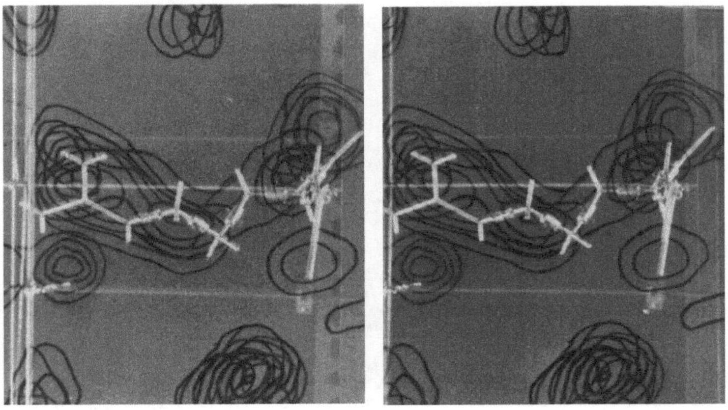

Fig. 21. Arg 254. This long and charged residue with the unique forking of the guanidium group is often well identifiable from the electron density map. However, owing to the long and flexible side chain this residue is sometimes only partially defined in the electron density maps. Arginine is frequently involved in hydrogen bond and salt bridge formation which makes it less flexible and also easier to identify

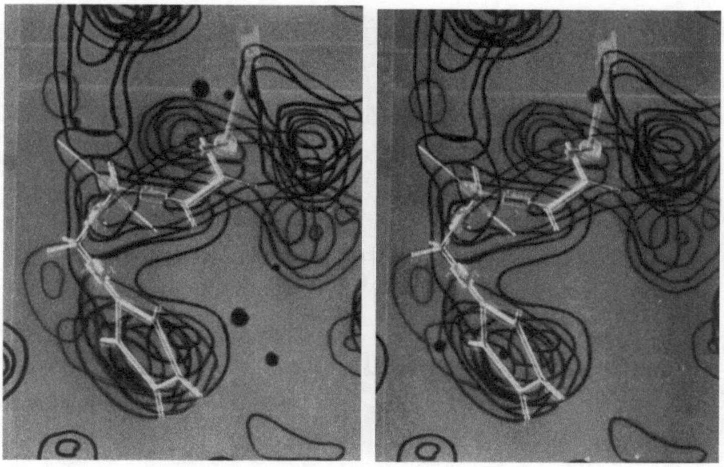

Fig. 22

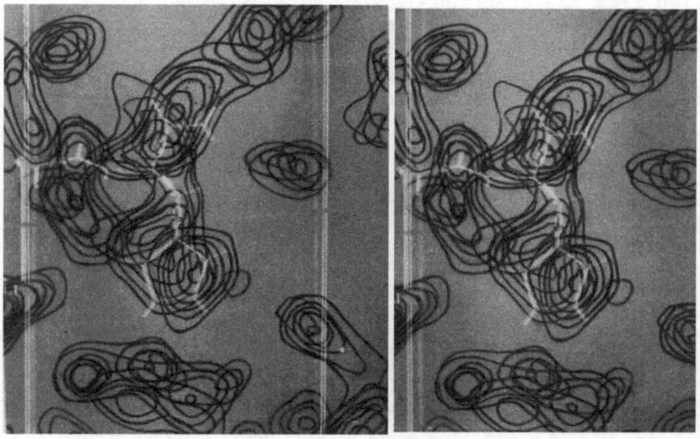

Fig. 23

Figs. 22 and 23. His 36 and 107. The smallest of the ring side chains is histidine. His is also a special amino acid residue often sought by protein chemists in the active sites of enzymes to explain the enzyme activity. The ring nature of the side chain associates this residue with a broad and flat electron density. When histidine is located on the surface of the protein accessible to the solvent the electron density may be a poor indicator of this residue. The hydrogen-bonding property of the two nitrogen atoms of histidine can stabilize the ring, giving rise to broad and continuous electron density to other residues as in Figure 23 above. His 107 is hydrogen-bonded to Glu 106 and Tyr 191 in the carbonic anhydrase molecule (330)

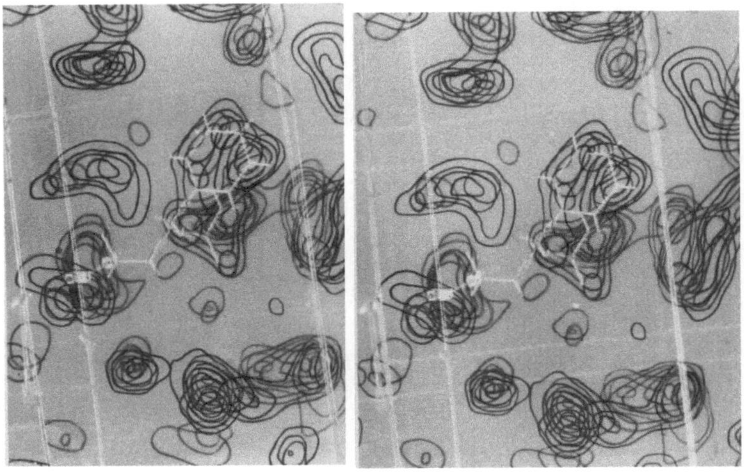

Fig. 24. Phe 179. Phenylalanine, with its aromatic ring, has often been easily identified from the electron density maps. A broad and rounded electron density occasionally with lower density in the middle is typical for phenylalanine. This hydrophobic residue is often buried in the interior of the molecule and is well stabilized by hydrophobic interactions. In a number of proteins, phenylalanine residues have been observed to form hydrophobic clusters (47, 131, 224, 330)

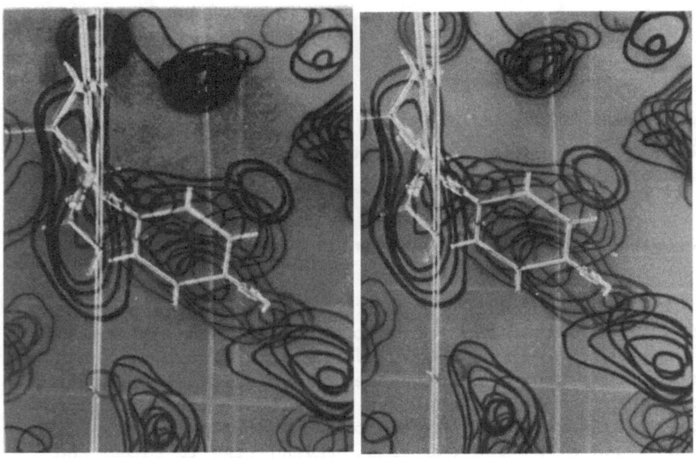

Fig. 25. Tyr 7. This aromatic residue with the hydroxyl group attached to the end of the ring, has also an associated broad electron density occasionally with lower density in the center of the ring. The presence of a protrusion at the end of the broad electron density is indicative of a tyrosine residue. The hydroxyl group can also participate in hydrogen-bonding, which stabilizes the ring. Obviously tyrosine and phenylalanine can be confused in some electron density maps

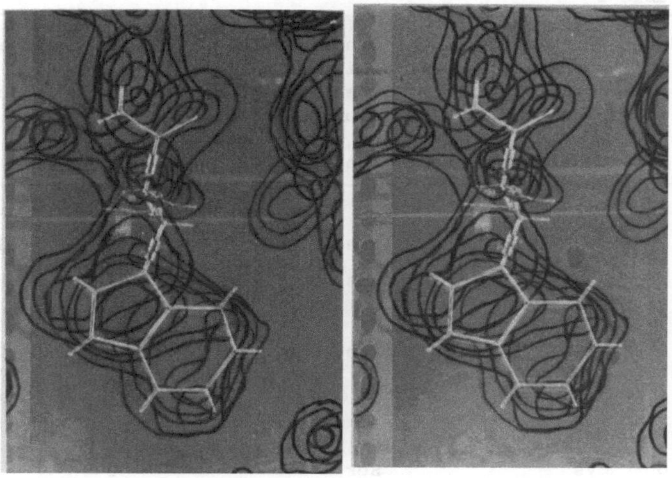

Fig. 26. Trp 192. By virtue of the two fused-rings, tryptophan is probably the residue most easily identified from electron density maps. The electron density of this residue is very broad and often shows a lower density in the middle of the six-membered ring

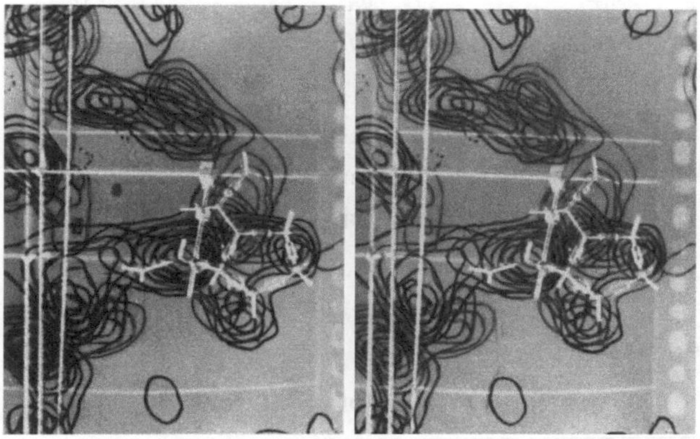

Fig. 27. Pro 195. The side chain of proline is bonded back to the main chain amide, making this residue part of the main chain. The electron density also reveals this characteristic by a broad density flowing together with the main chain. This has often made the identification of proline relatively easy from electron density maps. The peptide linkage in a protein molecule occurs almost exclusively in the trans conformation as determined from the electron density maps of almost all protein structures solved to date. However, the proline residue in the peptide linkage can take the cis conformation as well. This alternative conformation has been observed in a number of protein structures (9, 142, 205, 226, 330, 472, 474). The presence of the cis-proline may be inferred when the proline residue occurs in the third position of a reverse open turn (203, 226)

Table 2

No.	Amino acids in descending order of re-cognizability in the electron density maps	Other possible amino acid identifications from the shape of the electron density	Side chain environment in 5 proteins			
			Inside	Surface	Outside	Total
1	Trp		10	15	6	31
2	Tyr	Phe	10	34	23	67
3	Phe	His, Tyr	31	17	7	55
4	Pro		13	13	34	60
5	His	Phe, Glu, Gln	7	14	20	41
6	Gly		25	28	68	121
7	Met		7	1	2	10
8	Cys	Ser	3	7	4	24
9	Leu	Asp, Asn	45	19	9	73
10	Ile	Val, Thr	40	29	20	89
11	Val	Thr	39	32	24	95
12	Thr	Val	16	30	52	98
13	Arg	Lys	0	14	24	38
14	Glu	Gln, His	7	8	37	52
15	Gln	Glu, His	2	18	35	55
16	Asp	Asn, Leu	5	22	55	82
17	Asn	Asp, Leu	3	27	43	73
18	Ser		21	26	86	133
19	Ala		43	17	42	103
20	Lys		0	13	69	82
		Total	327	385	660	1372

Notes to Table 2:

1. The following proteins have been included in the calculation of the side chain environment: α-Chymotrypsin, carboxypeptidase, thermolysin, human carbonic anhydrase B and C. The values for α-chymotrypsin, carboxypeptidase and thermolysin are taken from text ref. 284.

2. The groupings in the recognizability column can vary from protein structure to protein structure.

Table 2 gives an account of the possible ambiguities that may arise between different side chains from electron density identifications and also the environment of different side chains in five protein molecules. The characteristic features of the electron densities of the side chains which can enable their identification are illustrated with stereo-photographs taken from the structure of human carbonic C. The sequence number (330) is also indicated. Well-defined side chain electron densities have been chosen to illustrate the characteristics of the individual amino acids.

E. Refinement of the Protein Structure

The identification of side chains from electron density maps is not a trivial operation and is possible only when the quality of the map is good and of sufficiently high resolution (2.8 to 2 Å). A certain acquaintance with the electron density map also gives the necessary inspiration and experience. The errors in the phase angles and the measured structure amplitudes often distribute over the whole map. In certain circumstances the errors can accumulate in particular regions of the electron density maps. This is often the case near heavy atom sites used in the phase-determination procedure. Such errors along with the flexibility in the side chains of protein molecules can produce a blurred image of the structure as seen from the electron density maps calculated on the basis of the multiple isomorphous replacement method.

In an effort to reduce the errors in the phase angles, protein crystallographers have tried to refine the protein structures and improve the quality and interpretability of the electron density maps (58, 110, 155, 202, 303, 457, 459). The coordinates (x_j, y_j, z_j) of the atoms that have been identified from the electron density map can be used in the following expression to calculate the structure factor of the reflection (h,k,l).

$$F_c(h,k,l) = \sum_{j=1}^{N} f_j \exp\left[2\pi i\ (h\ x_j + k\ y_j + l\ z_j)\right] \qquad (4)$$

where f_j is the scattering power of the j'th atom and the summation is over all the N atoms which have been identified with certainity from the electron density map. If all the atoms in the unit cell have been correctly identified the calculated structure amplitude $|F_c(h,k,l)|$ will be almost equal to that measured from the X-ray diffraction experiment $|F_o(h,k,l)|$. From the calculated structure factor the calculated phase angle $\alpha_c(h,k,l)$ is readily obtained and can be used either with $|F_o(h,k,l)|$ or the difference ($|F_o| - |F_c|$) between the observed can calculated structure amplitudes to calculate an observed electron density or a difference electron density map, respectively. The difference electron density has been shown to be useful in refining the protein structures determined by the isomorphous replacement method (58, 110, 155, 202, 303, 457, 459). The use of the refined phase angles generally improves the quality of the electron density map and increases the possibility of identifying missing atoms in the unit cell. Figures 28-31 illustrate Arg and Phe from the structure of pancreatic trypsin inhibitor as seen in the isomorphous and observed electron density maps, respectively. A criticism of this procedure is the fact that some of the wrongly placed atoms used in calculation of the structure factors $F_c(h,k,l)$ appear also in the observed electron density map. The refinement of a protein structure is not a trivial job either. However, the extremely interesting and useful results obtained in the case of a number of protein structure refinements such as more definitive identification of side chains (58, 110, 155, 202, 303, 457, 459) and the localization of atoms not included in the calculation certainly offsets this criticism. It is also worth mentioning in this context that the identification of a calcium ion in the structure of trypsin was achieved only (58) after the refinement of the protein structure.

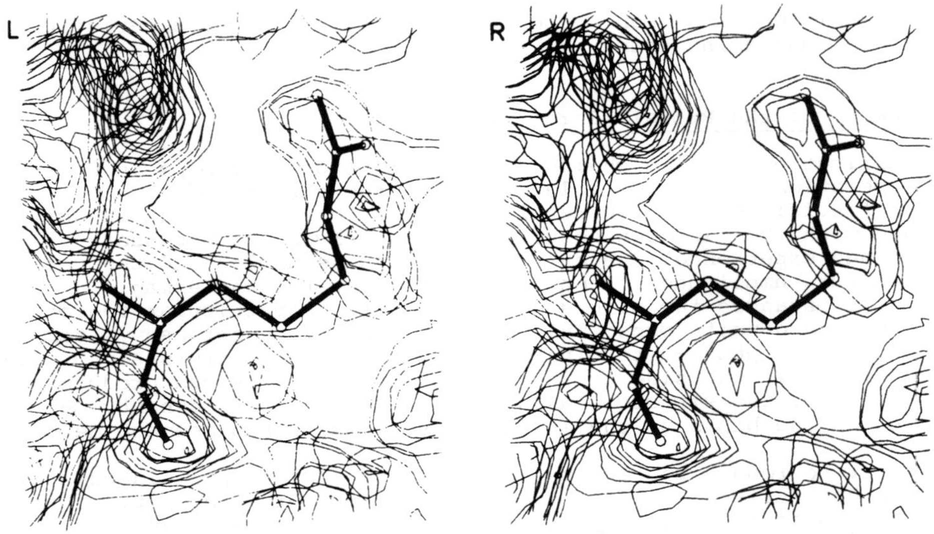

Fig.

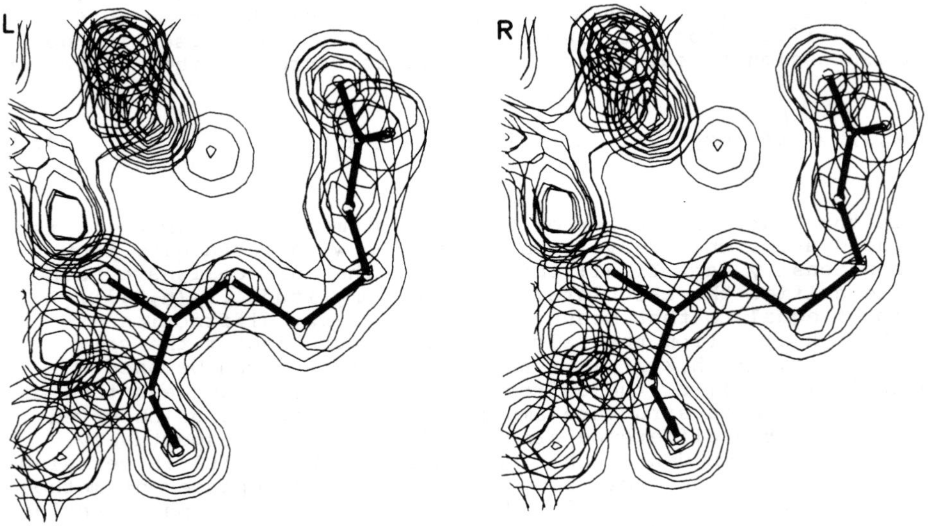

Fig. 2?

Figs. 28 and 29. Stereo pairs showing identical sections around Arg 53 of the 1.9 Å electron density map calculated with the isomorphous phases (top) and of the 1.5 Å resolution map calculated with refined phases (bottom). Contour lines are drawn at an interval of 0.3 e/A^3 starting from 0.3 e/A^3 in the 1.9 Å map and at an interval of 0.5 e/A^3 starting from 0.5 e/A^3 in the 1.5 Å map. The refined model of Arg 53 is drawn in both stereo-pairs (110) with the kind permission of Drs. J. Diesenhoffer, W. Steigemann and R. Huber

Figs. 30 and 31. Stereo pairs showing identical sections around Phe 45 of the 1.9 Å electron density map calculated with the isomorphous phases (top) and of the 1.5 Å resolution map calculated with refined phases (bottom). Contour lines are drawn at an interval of 0.3 e/A^3 starting from 0.3 e/A^3 in the 1.9 Å map and at an interval of 0.5 e/A^3 starting from 0.5 e/A^3 in the 1.5 Å map. The refined model of Phe 45 is drawn in both stereo-pairs (110) with the kind permission of Drs. J. Diesenhoffer, W. Steigemann and R. Huber

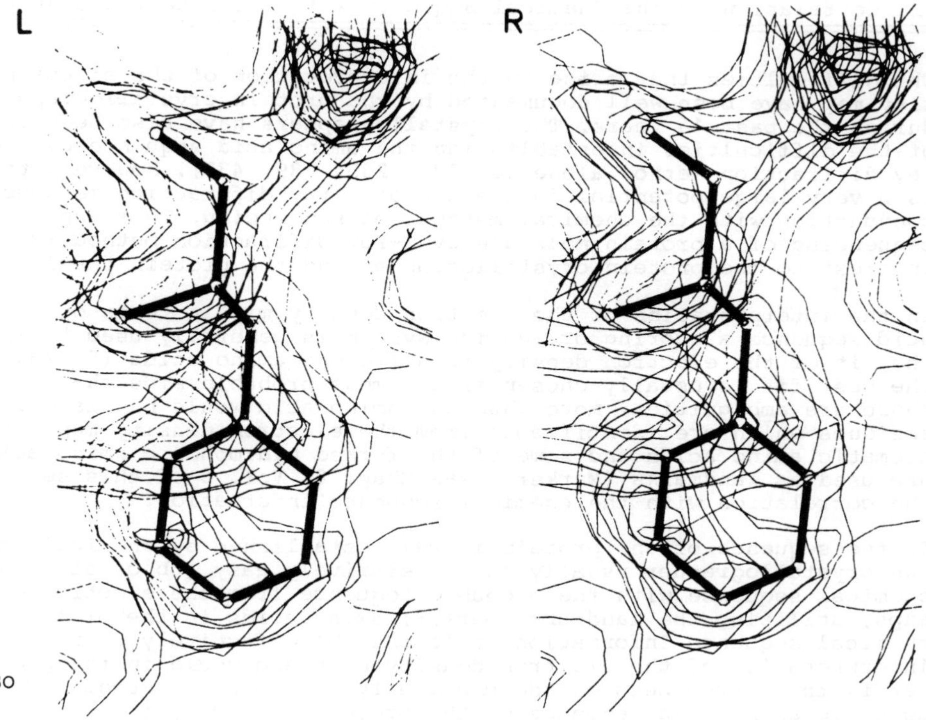

Fig. 30

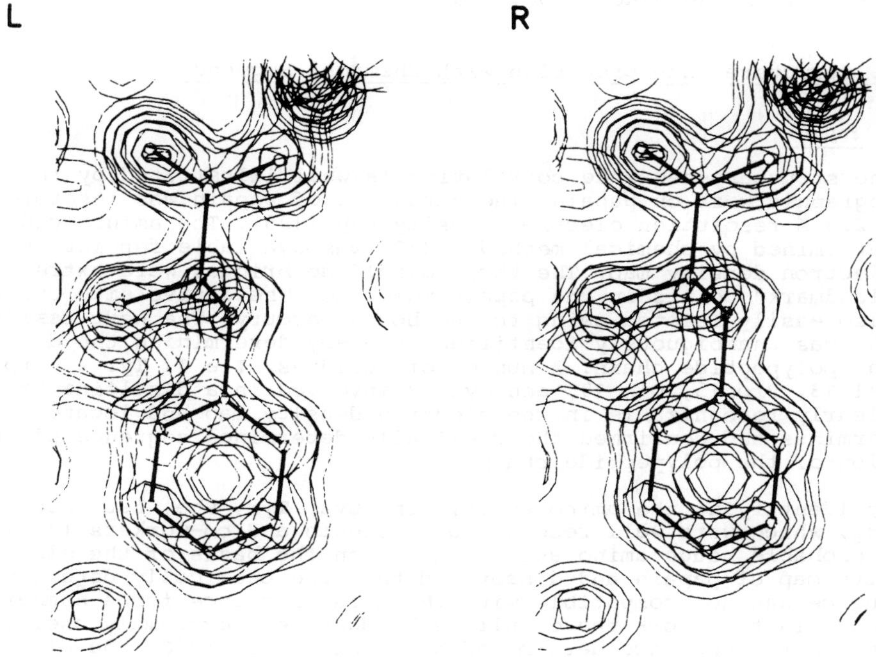

Fig. 31

VI. Correlation of the Chemical Sequence with the Electron Density Map

The difficulties that arise in the interpretation of the electron density map have been well documented by the protein crystallographers during the past 16 years. The crystallographers have also been aware of the difficulties in establishing the amino acid sequence by the X-ray diffraction method alone (5, 232, 265, 289, 435). However, there is a very large potential in the use of X-ray diffraction methods in conjunction with the chemical methods of sequencing. This aspect of th sequencing of a protein molecule by X-ray diffraction method is of gre interest to the protein crystallographer and the protein chemist alike

In the interpretation of the electron density map in terms of the amin acid sequence a scoring or ranking system is generally used to indicat the fit of the electron density to different amino acids (5, 232, 265) The best fit is usually chosen as the most probable side chain. Where there are ambiguities, more than one amino acid would be listed. The residues which are identifiable from the electron density map, such as aromatic amino acids and some of the forked hydrophobic amino acids are used as landmarks (Markers, see Chap. 8, Vol. 8 of this series) in the correlation with the chemical sequence information.

If the sequence of the protein is well established by chemical methods the crystallographer usually has a relatively easy job of aligning the chemical sequence with the probable sequence from the electron density maps, utilizing the landmark (marker) amino acids in the process. The chemical sequence information would thus be effectively used in the interpretation of the electron density maps and would in the process verify the correctness of the chemically determined sequence. When the complete amino acid sequence of the protein is known, a 3 Å or, favorable cases, even a 3.5 Å electron density map can be interpreted in terms of the sequence (409, 475).

A. Structure Interpretation with Chemical Sequence

1. Papain

The structure-sequence correlation is well illustrated by the crystallographic work on papain. The structure of papain was interpreted from a 2.8 Å resolution electron density map (125). The amino acid sequence determined by chemical methods (260) was available for interpreting th electron density map. The three disulfide bridges were extremely good "landmark" points in the papain molecule. The active site Cys 25 was also easily located owing to the bound mercury. The next residue, Trp 26, was unambiguously identified, thereby determining the direction of the polypeptide chain. A number of residues, Ile 1, Tyr 4, Trp 7, Arg Val 13, Pro 15, Gln 19, and Cys 22 involved in a disulfide bridge were clearly recognizable in the electron density map and located the amino terminus and confirmed the chemically determined sequence of this region of the polypeptide chain.

By identifying the amino acids, Trp, Cys, Tyr, Phe, Leu, Ile, Pro and Arg, which were well recognizable, Drenth and co-workers (125) propose a probable rough amino acid sequence on the basis of the electron density map of papain and discovered that the chemically determined sequence was not compatible with the X-ray sequence for a number of reasons. In this work the disulfide bridges were extremely useful. Accord ing to the sequence determined by Light et al. (260) the disulfide bridges should have been formed by Cys 22-Cys 159, Cys 43-Cys 152 and Cys 100-Cys 186. However, the electron density maps revealed Cys 22-Cys 63, Cys 56-Cys 95, Cys 153-Cys 200. It was also evident from the electron density maps that many residues were wrongly aligned and a number of residues had been entirely missed in the chemical sequence determination. The sequence that emerged from the electron density map gave the following arrangement of the chemically determined sequence: 1-28 with 13 extra residues including two Iles: 138-176, 31-137, 177-198 (Fig. 32). This revised amino acid sequence has since been chemically confirmed and corrected (207, 301), which provides an interesting illustration of the interaction in the sequence determination by X-ray crystallographic and chemical methods. However, without the correctly determined chemical sequence of the long peptides the X-ray sequencing would have been rather difficult.

2. Human Carbonic Anhydrase B

The electron density map of human carbonic anhydrase B (225, 226) could be interpreted easily with the chemical sequence (17, 262). The sequence reported by Andersson et al. (17) agreed generally well with the electron density map and ruled out the postulated absence of a tripeptide Ala 121 - His 122 - Trp 123, and a dipeptide Ala 135 - Ser 136 as well as an additional residue His 143 by Lin and Deutsch (262). This sequence error has since been corrected by Lin and Deutsch (262a).

Even when the chemical sequence is established by different scientists as in this case, the X-ray diffraction analysis might still serve to settle disputed sequences.

B. Structure Interpretation with Partial Sequence

For a number of protein structures, high resolution electron density maps had been computed before the complete chemical sequence became available. The main chain stretching and easily identified side chains could be established for these proteins. The different attempts to establish the sequence from electron density maps will be discussed below.

1. Myoglobin

Myoglobin is an extremely interesting example from the view point of sequencing from electron density maps. The complete amino acid sequence was not available when the high resolution structure was established (231). It was possible for Kendrew et al. (232) to arrive at a tentative sequence by correlating the amino acid identification from the electron density maps with the partial chemical sequence determined by Edmundson and Hirs (135). Table 3 illustrates this correlation. In tabulating the sequence Kendrew et al. (232) placed the sixteen tryptic peptides identified by Edmundson and Hirs (135) alongside the X-ray identification to derive the tentative sequence. Four tryptic peptides were assigned with less confidence. Two regions called core A and B were placed only on the basis of the X-ray identification and ordering of the peptide and the preliminary chemical ordering on the basis of chymotryptic digests.

Both the chemical and X-ray evidence were very preliminary and one would indeed expect that some revision would be necessary, as was suggested by Kendrew et al. (232). The final sequence in Table 3 confirms their opinion. The two core regions, where no chemical information was

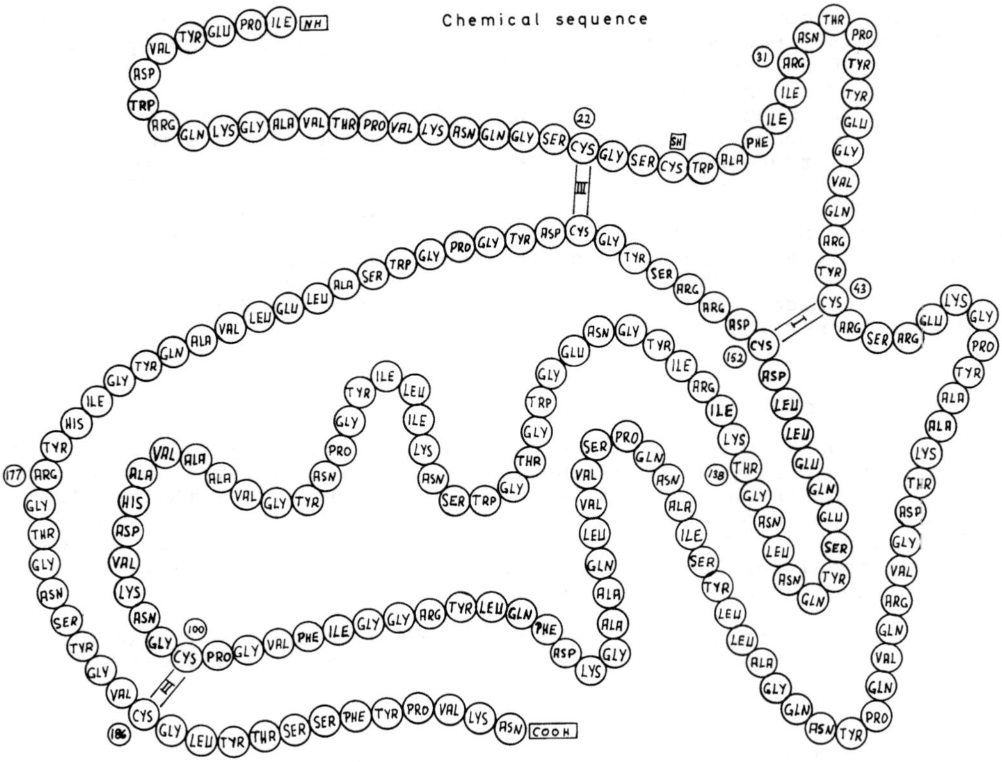

Fig. 32. (a) Tentative amino-acid sequence, according to Light et al. (260). (b) Amino-acid sequence as deduced from the electron density map

available, is of special interest from the point of view of sequencing only from electron density maps. The chemical evidence indicated one core peptide (44-46 residues) with a Val terminal group or several peptides, one of which contained a Val end group (135). The X-ray identification was not compatible with a single core peptide. Therefore Kendrew et al. (232) assigned the core residues to two peptides: core A consisting of about 32 residues and core B consisting of 13 residues This was since shown to be essentially correct (136). This superb example of tentative amino acid sequencing by a combination of the X-ray identification with partial chemical sequence analysis gives the true picture of the possibilities and limitations of this technique. It is clear from Table 3 that 60% correct assignement was achieved for the core peptide from X-ray diffraction analysis alone.

2. *Carboxypeptidase*

The high resolution structure of this enzyme was established before the complete chemical sequence was available. On the basis of the amin acid identification from the electron density maps and the tentative amino acid sequence already published (65, 97), Lipscomb et al. (265) arrived at a probable amino acid sequence. Table 4 shows that a large number of correct amino acid identifications could be made from the excellent electron density maps of carboxypeptidase available to

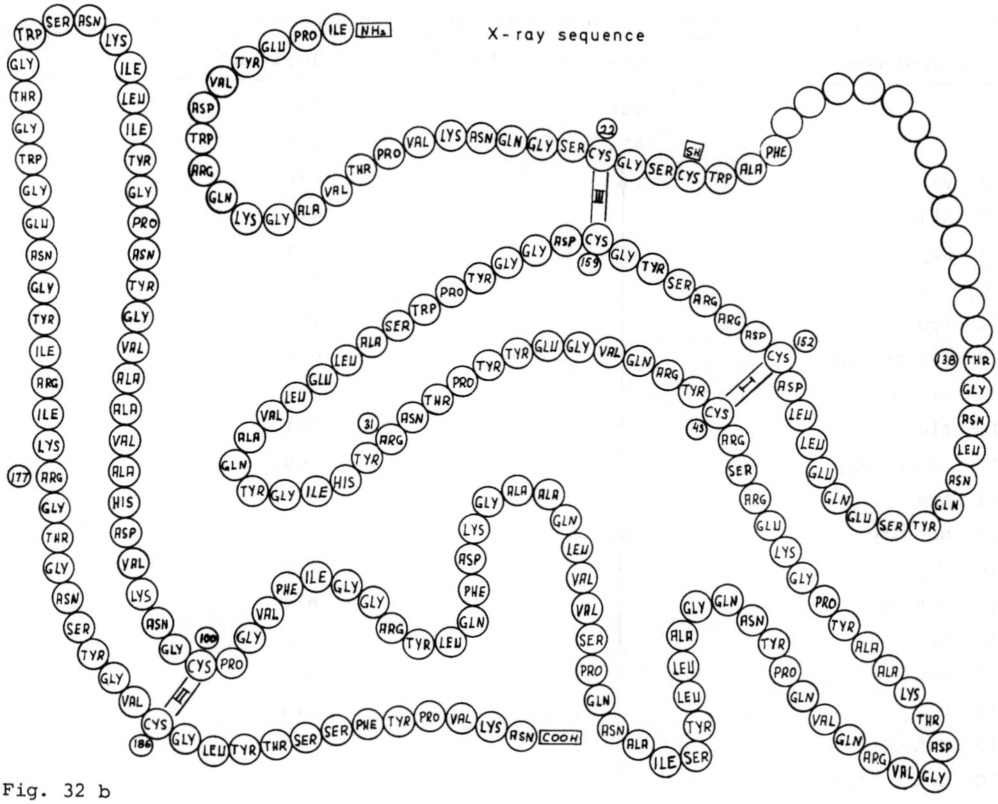

Fig. 32 b

Lipscomb and co-workers (264). It was also evident from these electron density maps that the cysteinyl residues 131 and 161 were involved in disulfide formation. This unexpected identification was confirmed in a chemical investigation by Bradshaw et al. (65). A number of functional residues were also identified from the electron density map, making it possible for Lipscomb et al. (264, 265) to suggest a tentative mechanism for the enzymic action of carboxypeptidase. 60% of the residues were correctly assigned from the electron density maps (264). In the highly ordered region covering residues 23-110 the X-ray sequencing was correct to a gratifying 75%.

3. *Human Carbonic Anhydrase C*

The high resolution structure of human carbonic anhydrase was determined (224, 261) when only fragmentary chemical sequence information was available (184). This sequence had been proposed by aligning the composition of different peptides obtained by cyanogen bromide cleavage, tryptic and chymotryptic hydrolysis, etc., with the partial sequence of the isoenzyme human carbonic anhydrase B (184). The single cysteinyl residue was assumed to be homologously located in the two sequences.

Five of 7 tryptophans, 6 of 8 tyrosines, 8 of 12 phenylalanines, 9 of 12 histidines, 14 of 18 prolines, the single methionine and a number of forked residues could be easily identified in the electron density

Table 3. X-ray sequence of myoglobin as compared to the chemical sequence

X-ray evidence[a]	Chemical evidence[a]	Sequence[b]
1[c]	Val	Val
2[c]	Ala	Leu
3 Val	↑	Ser
4 Ala		Glu
5 Gly		Gly
6 Glu		Glu
7 Trp		Trp
8 Ser or Ala		Gln
9 Glu or Ala		Leu
10 Ile		Val
11 Leu or Asp		Leu
12 Lys		His
13 Not Gly	Core A	Val
14 Trp		Trp
15 Not Gly		Ala
16 Leu		Lys
17 Leu		Val
18 Glu		Glu
19 Not Gly		Ala
20 Leu or Asp		Asp
21 Val or Thr		Val
22 Ala		Ala
23 Gly		Gly
24 His		His
25 Gly or Ala		Gly
26 Lys		Gln
27 Leu or Asp		Asp
28 Thr or Val or Asp		Ile
29 Leu		Leu
30 Ile		Ile
31 Ser		Arg
32 Leu		Leu
33 Phe		Phe
34 Thr or Lys or Ser	↓	Lys
35 Ser or Thr	Ser	Ser
36 His	His	His
37 Pro	Pro	Pro
38 Glu	Glu	Glu
39 Thr	Thr	Thr

Table 3 (continued)

X-ray evidence[a]		Chemical evidence[a]	Sequence[b]
40	Leu	Leu	Leu
41	Ser or Thr	Glu	Glu
42	Lys	Lys	Lys
43	Phe	Phe	Phe
44	Asp	Asp	Asp
45	Arg	Arg	Arg
46	Phe	Phe	Phe
47	Lys	Lys	Lys
48	His	His	His
49	Leu or Asp	Leu	Leu
50	Lys	Lys	Lys
51	Thr	Thr	Thr
52	Glu	Glu	Glu
53	Ala	Ala	Ala
54	Glu	Glu	Glu
55	Met	Met	Met
56	Lys	Lys	Lys
57	Ala	Ala	Ala
58	Ser	Ser	Ser
59	Glu	Glu	Glu
60	Asp or Leu	Leu	Asp
61	Leu or Asp	Asp	Leu
62	Lys	Lys	Lys
63	Val or Thr or Ile	Val	Lys
64	His or Glu	Glu	His
65	Gly	Ala	Gly
66	Ile	Ile	Val
67	Glu or Ser	Glu	Thr
68	Val or Thr	Val	Val
69	Asp or Lys	Asp	Leu
70	Ser or Thr or Ala	His	Thr
71	Ala or Gly	Gly	Ala
72	Leu	Leu	Leu
73	Gly	Gly	Gly
74	Ala	Ala	Ala
75	Ile or Val or Thr	Ile	Ile
76	Asp or Leu	Asp	Leu
77	Arg or Lys	Arg	Lys
78	Lys	Lys	Lys

Table 3 (continued)

X-ray evidence[a]		Chemical evidence[a]	Sequence[b]
79	Lys	Lys	Lys
80	Gly	Gly	Gly
81	Leu or Asp or His	Leu	His
82	His	His	His
83	Glu or Ala	Glu	Glu
84	Not Gly	Leu	Ala
85	Glu	Glu	Glu
86	Glu or His	Glu	Leu
87	Ala	Ala	Lys
88	Pro	Pro	Pro
89	Thr or Val or Ser	Thr	Leu
90	Ala or Gly	Ala	Ala
91	His or Asp	His	Gln
92	Ser or Ala	Ser	Ser
93	His	His	His
94	Ala	Ala	Ala
95	(Not Ala, or Gly) Lys ? β-forking	Lys	Thr
96	Not Gly	Leu	Lys
97	Phe or His	Phe	His
98	(Not Ala or Gly) β-forking, Lys ?	Lys	Lys
99	Ile	Ile	Ile
100	Pro	Pro	Pro
101	Ile	Ile	Ile
102	Lys	Lys	Lys
103	Tyr	↑	Tyr
104	(Not Ala, Gly) or β-forking		Leu
105	Glu	Core B	Glu
106	Phe, His		Phe
107	Leu, Asp		Ile
108	Ser		Ser
109	(Not Gly, Ala)		Glu
110	Ala		Ala
111	Val or Thr, Ile		Ile
112	Ile, Val or Thr		Ile
113	His		His
114	Val or Thr		Val
115	Arg, Leu, Asp, Lys	↓	Leu

Table 3 (continued)

X-ray evidence[a]	Chemical evidence[a]	Sequence[b]
116 Ala, Asp, Glu	Ala	His
117 Ala, Thr, Val, Ser	Thr	Ser
118 Lys	Lys	Arg
119 His	His	His
120 Ala, Glu	Asp	Pro
121 Gly, Asp	Asp	Gly
122 Glu	Glu	Asn
123 Phe	Phe	Phe
124 Gly	Gly	Gly
125 Ala	Ala	Ala
126 Pro	Pro	Asp
127 Ala	Ala	Ala
128 Asp, Ser, β-forking	Asp	Gln
129 Gly, Lys	Gly	Gly
130 Ala	Ala	Ala
131 Met	Met	Met
132 Gly, β-forking	Gly	Asn
133 Lys	Lys	Lys
134 Ala	Ala	Ala
135 Leu	Leu	Leu
136 Glu	Glu	Glu
137 Leu	Leu	Leu
138 Phe	Phe	Phe
139 Arg	Arg	Arg
140 Lys, Ser	Lys	Lys
141 Asp	Asp	Asp
142 Ile	Ile	Ile
143 Ala	Ala	Ala
144 Ser or Ala	Ala	Ala
145 Lys or Ile	Lys	Lys
146 Tyr	Tyr	Tyr
147 (Not Gly, Ala β-forking) Lys ?	Lys	Lys
148 Glu	Glu	Glu
149 Leu, Lys	Leu	Leu
150 Gly	Gly	Gly
151 Tyr	Tyr	Tyr
152 Gly	Gly	Gln
153	Glu	Gly

Notes to Table 3:

[a]Adapted from Kendrew et al. (1961), Nature 190, 666.

[b]Edmundson, A.B. (1965), Nature 205, 883.

[c]Authors alignment.

Notes to Table 4:

Table showing sequence of first 103 residues of carboxypeptidase, indicating large percentage of correct identifications that could be made from the electron density map. Column 2 gives the initial identifications made from the map. Column 3 gives the final identifications from the combination of chemical and X-ray methods. Parentheses in column 2 enclose secondary choices, commas separate equal choices; parentheses in column 3 indicate discrepancies remaining after the chemical and X-ray sequences were compared; these are discussed in the footnotes. Where column 2 and column 3 differ, and there is no note, it is to be assumed the choice in column 3 was found to be consistent with the map. A + in the notes column indicates that a correct but uncertain chemical identification was confirmed by the X-ray data. A * indicates that no identification could be made from the X-ray data alone, but a guess is given in column 2.

Abbreviations used: Glx = glutamic acid or glutamine, Asx = aspartic acid or asparagine

The following footnotes are numbered according to the residue to which they refer.

1. Residues 1-22 were built directly from previously available chemical sequence information. Therefore, independent X-ray identifications were not attempted.

18. At one point in the X-ray analysis, it was felt that No. 18 must be Val. However, later repositioning of the backbone allowed Ile to fit.

26. At one point in the X-ray analysis Ile appeared to be required here.

27. The electron density map cannot support Ala at No. 27.

29. X-ray identification of His 29 was based primarily on heavy metal binding rather than electron density.

33. The map, although weak at this position, cannot be reconciled with Val.

41. The map shows a branched residue here which cannot be reconciled with Ser.

48. This and another Tyr were for a time confused with arginine, which can be folded into a conformation exactly like Tyr, with one ring atom missing. Such a conformation is unlikely, however, for arginine, and tyrosine should probably be chosen for all side-chains giving this appearance.

52. This phenylalanine has a pentagonal appearance, but may be identified as Phe from the hydrophobic environment.

62. After several examinations, the map still supports Leu rather than Ile.

68.
69. The X-ray result that No. 69 is the histidine ligand to zinc enabled the correct chemical identification of 68 and 69 to be made. The X-ray guess Leu for No. 68 was an early error — Ile actually fits the density better.

85. Correct identification of 85 was hindered by its being curled up rather than fully extended.

88.
89. The strong X-ray identification of 89 as Asx enabled the correct chemical identification of 88 and 89 to be made (there must be one each of Glx and Asx in this region). The map at No. 88 was consistent with Asx or Glx.

Table 4. Sequence of residues 1-103 for CPA

Residue No.	Model-building X-ray identification	Final: chemical and X-ray	Footnote
1	Ala	Ala	*, 1
2	Arg	Arg	
3	Ser	Ser	
4	Thr	Thr	
5	Asx	Asn	*
6	Thr	Thr	
7	Phe	Phe	
8	Asx	Asn	
9	Tyr	Tyr	
10	Ala	Ala	
11	Thr	Thr	
12	Tyr	Tyr	
13	His	His	
14	Thr	Thr	
15	Leu	Leu	
16	Asx	Asp	
17	Glx	Glu	
18	Ile	Ile	18
19	Tyr	Tyr	
20	Asx	Asp	
21	Phe	Phe	
22	Met	Met	
23	Asx	Asp	
24	Leu (Ile)	Leu	
25	Leu, Ile, Asx	Leu	
26	Ile (Val)	Val	26
27	Gly (Ala)	Ala (Gly)	27
28	Glx	Gln	
29	His	His	29
30	Pro	Pro	
31	Glu ??	Glu	*
32	Leu (His, Asx)	Leu	+
33	Leu, Asx (Ile)	Val (Leu)	33
34	Ser, Asx, etc.	Ser	
35	Lys	Lys	+
36	Leu	Leu	*
37	Glx, Lys	Gln	
38	Ile	Ile	

Table 4 (continued)

Residue No.	Model-building X-ray identification	Final: chemical and X-ray	Footnote
39	Gly	Gly	
40	Arg	Arg	
41	Thr (Ser)	Ser (Thr)	41
42	Tyr (His)	Tyr	
43	Glx	Gln	
44	Gly	Gly	
45	Arg	Arg	+
46	Pro, Thr	Pro	
47	Ile ?	Ile	
48	Tyr	Tyr	48
49	Ser, Val	Val	
50	Leu	Leu	+
51	Lys, Glx	Lys	
52	His, Phe	Phe	52
53	Ser	Ser	
54	Thr (Ser, Ile)	Thr	*
55	Gly	Gly	
56	Gly	Gly	
57	Gly, Ser, etc.	Ser	*
58	Asp, Ala	Asn	
59	Arg	Arg	
60	Pro ?	Pro	+
61	Ala ? (Gly)	Ala	+
62	Leu, Ile	Ile (Leu)	62
63	Trp	Trp	+
64	Ile	Ile	+
65	Asx	Asp	
66	Leu	Leu	+
67	Gly	Gly	
68	Leu	Ile	+, 68
69	His	His	+, 69
70	Ser	Ser	
71	Arg	Arg	
72	Glx	Glu	
73	Trp	Trp	+
74	Ile ? (Ser)	Ile	
75	Thr	Thr	*
76	Glx	Gln	+

Table 4 (continued)

Residue No.	Model-building X-ray identification	Final: chemical and X-ray	Footnote
77	Ala	Ala	
78	Thr, Ile	Thr	
79	Gly	Gly	+
80	Val	Val	+
81	Trp	Trp	+
82	Phe	Phe	
83	Ala	Ala	
84	Lys	Lys	+
85	His, Lys	Lys	85
86	Phe	Phe	+
87	Thr	Thr	
88	Asx. Glx, Lys	Glx	*, 88
89	Asx	Asx	+, 89
90	Tyr	Tyr	+
91	Gly	Gly	+
92	Asx	Gln	
93	Glx ?	Asn	
94	Pro	Pro	+
95	Ser	Ser	+
96	Phe	Phe	+
97	Thr	Thr	+
98	Ala	Ala	
99	Val, Ile	Ile	
100	Leu, Asx	Leu	
101	Asx (Glx)	Asp	+
102	Ser	Ser	+
103	Met	Met	

maps (unpublished results). The zinc ligands were identified as histidyl residues 94, 96 and 118 (224, 261) which has subsequently been confirmed (185). An initial attempt to utilize the sequence information for the B enzyme and the peptide composition data for the C enzyme to arrive at an X-ray sequence proved to be difficult and was not pursued for a number of reasons. The cysteinyl residue was located on the basis of the location of the bound mercury (434). A stretch of the peptide chain just before the cysteinyl residue had to be built in an invisible electron density region, even though the continuation of the electron density indicated otherwise, in order to accomodate the partial sequence information available based on homology. It was not always possible to correlate the composition data of the different peptides with the electron density identification. Subsequently, when the chemical sequence was fully established it was easier to correlate the structure

Table 5. Amino acid sequence for *C. pasteurianum* rubredoxin

No. of R group	Ident. from 2 Å map[a]	Confi- dence	Ident. from refinement	Ident. from chem.[b]
1	Met	3		fMet
2	Leu	1	Lys	Lys
3	Lys	3		Lys
4	Tyr	3		Tyr
5	Thr	2		Thr
6	Cys	3		Cys
7	Thr	2		Thr
8	Val	2		Val
9	Cys	2		Cys
10	Gly	3		Gly
11	Tyr	3		Tyr
12	Asx	3	?	Ile
13	Tyr	3		Tyr
14	Thr	2	Asx	Asp
15	Pro	3		Pro
16	Asx	2	?	Glu
17	Leu	1	Prob. Asx	Asp
18	Gly	3		Gly
19	Asx	3		Asp
20	Pro	3		Pro
21	Val	2	?	Asp
22	Asx	3		Asp
23	Gly	3		Gly
24	Ile	2	Val	Val
25	Asx	2		Asn
26	Pro	3		Pro
27	Gly	3		Gly
28	Val	3	Thr	Thr
29	Glx	2	Prob. Asx	Asp
30	Phe	3		Phe
31	Lys	2		Lys
32	Asx	3		Asp
33	Ile	2		Ile
34	Pro	3		Pro
35	Asx	3		Asp
36	Asx	2		Asp
37	Trp	3		Trp
38	Val	2		Val

Table 5 (continued)

No. of R group	Ident. from 2 Å map[a]	Confidence	Ident. from refinement	Ident. from chem.[b]
39	Cys	3		Cys
40	Pro	3		Pro
41	Val	2	Leu	Leu
42	Cys	3		Cys
43	Gly	3		Gly
44	Ile	2	Val	Val
45	Gly	3		Gly
46	Lys	3		Lys
47	Asx	2		Asp
48	Glx	3		Glu
49	Phe	3		Phe
50	Glx	2		Glu
51	Glx	3		Glu
52	?	1	Val	Val
53	Asx	3	Prob. Glx	Glu
54	?	1	?	Glu

[a]For explanation of confidence levels, see text.

[b]McCarthy (292).

and sequence. The loop with invisible electron density vanished when the sequence established that the cysteinyl residue is not homologously located and the peptide chain could then be built as indicated by the electron density map (330). An additional loop was included to accommodate the completed sequence in a discontinuous albeit strong electron density region on the surface of the molecule. All the unexplained electron densities could thus be correlated very well with the sequence.

4. Rubredoxin

Only the chemical composition was known for this small 54 amino acid protein when the high resolution structure was determined by Herriot et al. (187). Table 5 presents the comparison of the X-ray sequence with the chemical sequence (186). The accompanying figure of Herriot et al. (186) gives a very useful correlation of the 2 Å electron density map with the amino acid sequence (Fig. 33). On the basis of the excellent electron density map of Rubredoxin, Herriot et al. (187) achieved 75% correct assignment of the sequence from X-ray diffraction work alone. Herriot et al. (186) have also given a lucid discussion of their amino acid sequencing from the electron density maps.

5. Thermolysin

The high resolution structure and the chemical sequence were completed simultaneously for thermolysin (97, 288). However, Colman et al. (97)

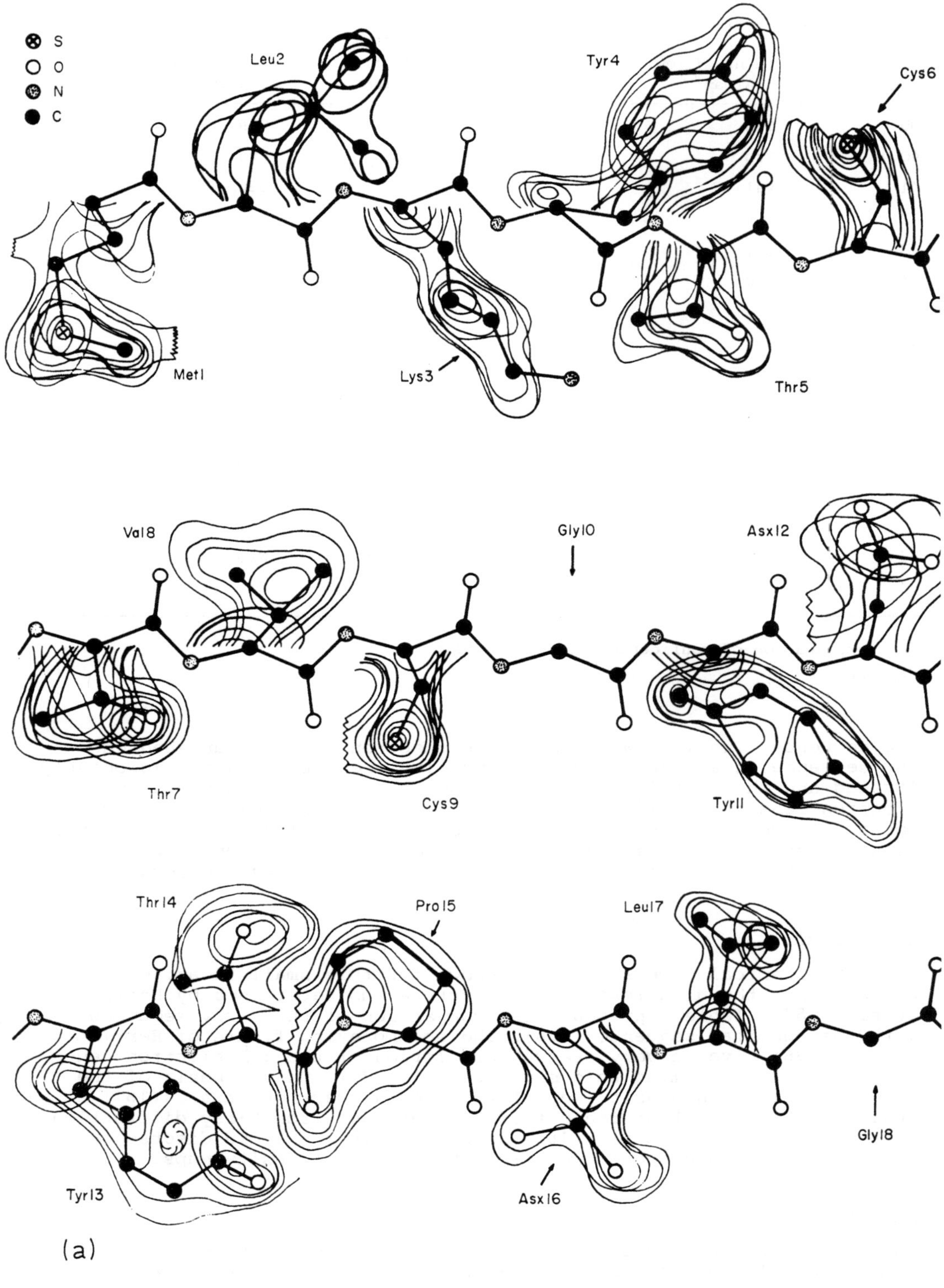

S
O
N
C
Leu2
Tyr4
Cys6
Met1
Lys3
Thr5
Val8
Gly10
Asx12
Thr7
Cys9
Tyr11
Thr14
Pro15
Leu17
Tyr13
Asx16
Gly18
(a)

Fig. 33a

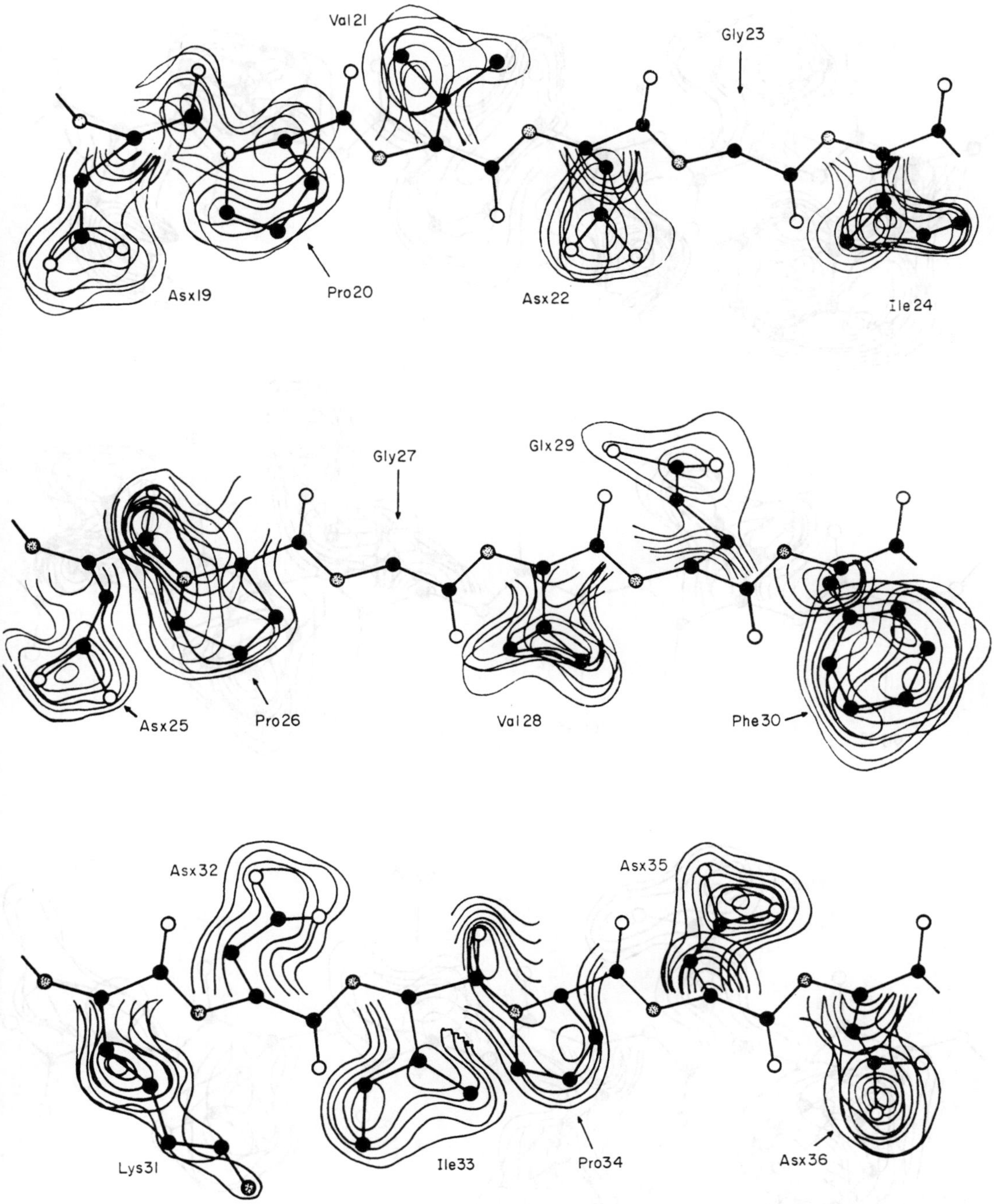

(b)

Fig. 33a-c. Electron density map with the amino acid sequence of Rubredoxin

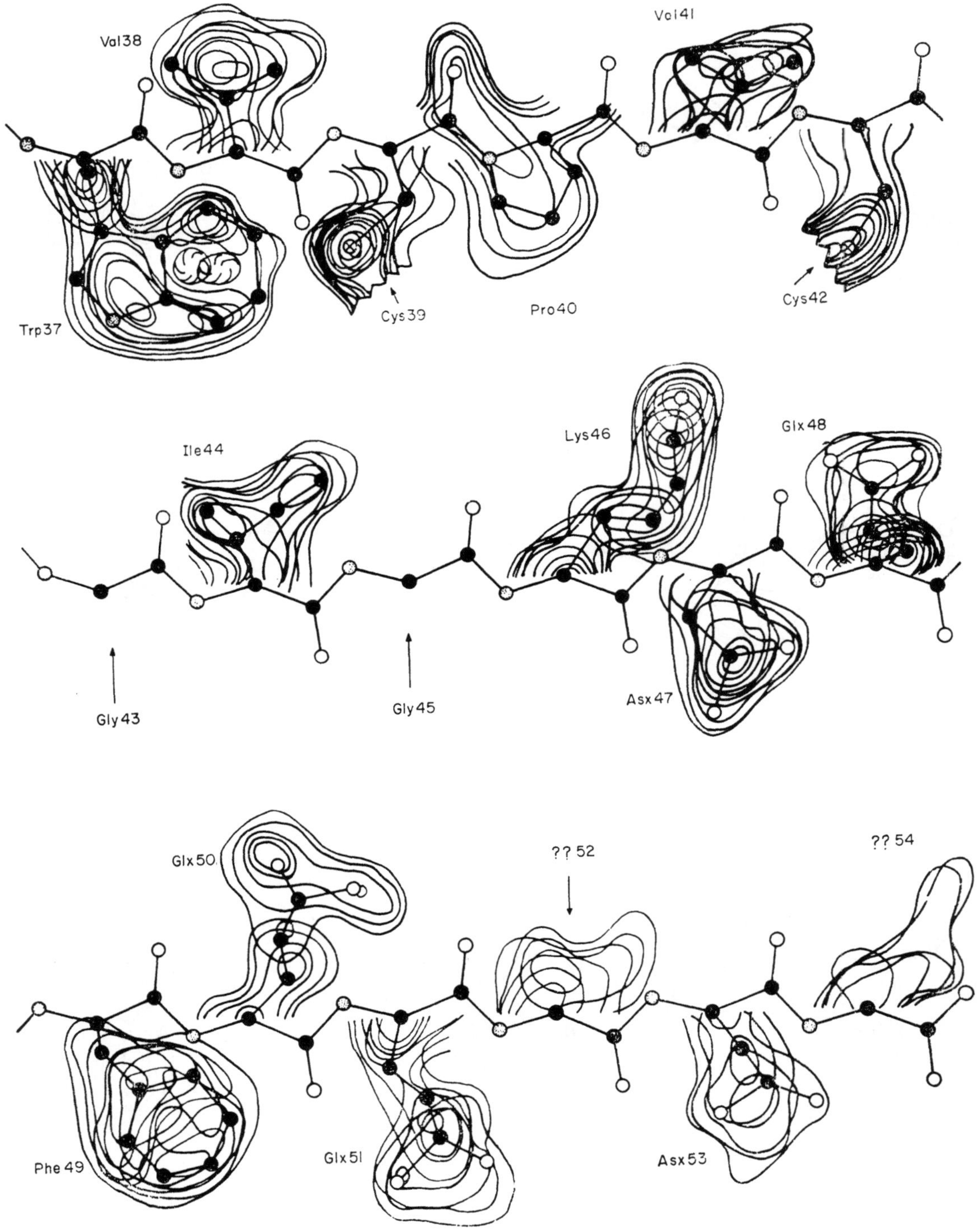
Val38
Val41
Trp37
Cys39
Pro40
Cys42
Ile44
Lys46
Glx48
Gly43
Gly45
Asx47
Glx50
?? 52
?? 54
Phe49
Glx51
Asx53
(c)

Fig. 33c

Table 6. Comparison of the chemical sequence with the X-ray sequence of Thermolysin. The X-ray sequence that differed from the chemical sequence is given below the chemical sequence

```
                 5                  10                  15                  20
Ile-Thr-Gly-Thr-Ser-Thr-Val-Gly-Val-Gly-Arg-Gly-Val-Leu-Gly-Asp-Gln-Lys-Asn-Ile-
            Ala                  Asp    Ile     Ile Val         Off Gly

                25                  30                  35                  40
Asn-Thr-Thr-Tyr-Ser-Thr-Tyr-Tyr-Tyr-Leu-Gln-Asp-Asn-Thr-Arg-Gly-Asp-Gly-Ile-Phe-
Ile Val Ser     Gly Val                     Gln     Gly             Ala

                45                  50                  55                  60
Thr-Tyr-Asp-Ala-Lys-Tyr-Arg-Thr-Thr-Leu-Pro-Gly-Ser-Leu-Trp-Ala-Asp-Ala-Asp-Asn-
Ser     Gly             Lys         Ile                *     ?      Gly  ?

                65                  70                  75                  80
Gln-Phe-Phe-Ala-Ser-Tyr-Asp-Ala-Pro-Ala-Val-Asp-Ala-His-Tyr-Tyr-Ala-Gly-Val-Thr-
Lys     Off     Val Phe Ser     Lys     Leu                                 Ser

                85                  90                  95                  100
Tyr-Asp-Tyr-Tyr-Lys-Asn-Val-His-Asn-Arg-Leu-Ser-Tyr-Asp-Gly-Asn-Asn-Ala-Ala-Ile-
    Glu Phe     Asn Thr     Gln Thr     Thr         Ser     Lys Thr

                105                 110                 115                 120
Arg-Ser-Ser-Val-His-Tyr-Ser-Gln-Gly-Tyr-Asn-Asn-Ala-Phe-Trp-Asn-Gly-Ser-Glu-Met-
    Asn Thr     Glu     Thr Arg Ala                 Glu     Pro             Leu

                125                 130                 135                 140
Val-Thr-Gly-Asp-Gly-Asp-Gly-Gln-Thr-Phe-Ile-Pro-Leu-Ser-Gly-Gly-Ile-Asp-Val-Val-
Thr                     Ile Thr Asp Arg         Ile

                145                 150                 155                 160
Ala-His-Glu-Leu-Thr-His-Ala-Val-Thr-Asp-Tyr-Thr-Ala-Gly-Leu-Ile-Tyr-Gln-Asn-Glu-
    Met Gly Val                 Asn         Gly Pro Ser     Thr     Leu

                165                 170                 175                 180
Ser-Gly-Ala-Ile-Asp-Glu-Ala-Ile-Ser-Asp-Ile-Phe-Gly-Thr-Leu-Val-Glu-Phe-Tyr-Ala-
                    Asp     Leu         Thr His             Ser         Gly

                185                 190                 195                 200
Asn-Lys-Asn-Pro-Asp-Trp-Glu-Ile-Gly-Glu-Asp-Val-Tyr-Thr-Pro-Gly-Ile-Ser-Gly-Asp-
Thr Tly Lys Ser Gly Phe Ser Leu      ?              ?          Val Lys Ala

                205                 210                 215                 220
Ser-Leu-Arg-Ser-Met-Ser-Asp-Pro-Ala-Lys-Tyr-Gly-Asp-Pro-Asp-His-Tyr-Ser-Lys-Arg-
Ala Ile     Asp Leu Lys Thr Ala Thr Ser Gln     Glu Asn Gln         Ala

                225                 230                 235                 240
Tyr-Thr-Gly-Thr-Gln-Asp-Asn-Gly-Gly-Val-His-Ile-Asn-Ser-Gly-Ile-Ile-Asn-Lys-Ala-
    Val     Val Lys Thr     Ser         Gly Val         Ser Asn Leu Thr Gly

                245                 250                 255                 260
Ala-Tyr-Leu-Ile-Ser-Gln-Gly-Gly-Thr-His-Tyr-Gly-Val-Ser-Val-Val-Gly-Ile-Gly-Arg-
        Asp Leu     Asp Ala     Asp Met Phe     Thr     Ser Pro             Trp

                265                 270                 275                 280
Asp-Lys-Leu-Gly-Lys-Ile-Phe-Tyr-Arg-Ala-Leu-Thr-Gln-Tyr-Leu-Thr-Pro-Thr-Ser-Asn-
    Arg      -          Tyr     Lys     Val Asn Ser     Ile Ala      ?  Ile His

                285                 290                 295                 300
Phe-Ser-Gln-Leu-Arg-Ala-Ala-Ala-Val-Gln-Ser-Ala-Thr-Asp-Leu-Tyr-Gly-Ser-Thr-Ser-
        Arg Asp             Ser         Ala Gly Gly Glu Glu     Ala Lys    *Thr

                305                 310                 315
Glx-Glx-Val-Ala-Ser-Val-Lys-Gln-Ala-Phe-Asp-Ala-Val-Gly-Val-Lys
Ile Asp Thr                         Tyr                     Gly
```

made an attempt to determine the sequence from the electron density map alone and to correlate this with the sequence determined chemically by Titani et al. (437). Table 6 illustrates this interesting correlation where they achieved 53% correct assignment from a quick X-ray identification alone. With a more careful interpretation and model building, a somewhat better X-ray sequence should have been possible, considering the good quality of the electron density map of thermolysin (97).

6. Lactate Dehydrogenase

The chemical sequences of only a few peptides were known when the high resolution structure of lactate dehydrogenase (4) was interpreted. These peptides could be located easily in the electron density maps and have been verified by completion of the chemical sequence (430). The sequences of peptides as they became available were communicated to the crystallographers and compared with the electron density maps (5). The possible identity of the side chain density was plotted on a grade scale of A to E in a matrix, X-ray matrix, consisting of the 20 amino acids on the horizontal direction and the peptide sequence number in the vertical direction. A similar matrix was constructed on a transparent sheet with the chemical sequence of a peptide in the vertical direction. The correct amino acid was indicated by a circle in this matrix which was in the vertical direction over the X-ray matrix. The best fit between the two was chosen as the correct location. Table 7 shows such a matrix system with about 50% correct fit for the peptides 44-76.

7. Cytochrome c_{550}

The structure-sequence cooperation is extremely well illustrated by the work of Timkovich et al. (435) on cytochrome c_{550} from Paracoccus denitrificans (Table 8). When the 2.5 Å structure of cytochrome c_{550} was completed, the amino acid sequence determination was initiated. The tryptic peptides that were sequenced by standard chemical methods were then ordered on the basis of the electron density maps treating the peptides as pieces in a jig-saw puzzle. The correctness of this alignment was checked on the basis of the amino acid composition of peptides obtained by a chymotryptic digest. The final sequence arrived at on the basis of this excellent cooperation is illustrated in Table 7. Even if the sequence might eventually be revised by chemical methods, the structure and sequence proved to be useful in establishing the sequence homology among tuna cytochrome c, *Rhodomicrobium vanielli* c_2, *Rhodospirillum rubrum* c_2, *Rhodopseudomonas palusteris* c_2, *Rhodopseudomonas spheroides* c_2, *Rhodopseudomonas capsulata* c_2 and *Paracoccus dentrificans* c_{550} (285).

VII. Conclusion

The success rate for sequence determination from X-ray diffraction analysis alone is 50-60% when the electron density maps are of high quality. It is evident from the above examples that a somewhat higher success rate is achieved when partial chemical sequence data are availalbe. It is not always possible to distinguish between acids and amides from electron density maps, which could mean that the success rate is somewhat overestimated above. It has been appropriately pointed out by Lipscomb et al. (265) that: "In spite of the possibilities of development of general environmental and conformational principles to assist these identifications and of objective fits of electron density to residues, it seems extremely unlikely that any X-ray study alone will produce a correct, or even a nearly correct, sequence. This method is intrinsically limited by disorder, especially of side chains which extend into solution in the protein crystal. In addition, some nearly isosteric pairs, for example Val and Thr, are especially difficult to distinguish."

The aim of this article has been to examine the potential for X-ray identification of side chains leading to a tentative sequence derived from this basis and which can be correlated with the chemically determined sequence. The obvious usefulness of this procedure lies especially in:

1. Cross-checking the correctness of the structure and the chemical sequence when available.
2. Aligning and ordering sequenced tryptic peptides or, as in the case of concanavalin A (131), ordering sequenced cyanogen-bromide fragments.
3. Detecting errors in the chemical sequence, such as missed or misaligned peptides, by X-ray diffraction methods.
4. Identifying important residues in the active site of an enzyme by X-ray diffraction methods and utilizing this information in elucidating their implication in the enzyme mechanism.

But this usefulness could be achieved only when the available partial chemical sequence is very nearly correct. It may not always be feasible to arrive at the sequence of peptides only on the basis of the amino acid composition data and electron density map. However, it seems clear from the example of cytochrome c_{550} (435) that, if the chemist and the crystallographer cooperate throughout the process of sequence and structure determination work, time and effort can be saved, while gaining very useful functional information of the protein arrived at easily.

Acknowledgments. I would like to thank Professor B. Strandberg, Drs. A. Liljas, I. Vaara, D. Eaker and H. Cid-Dresdner for very useful criticism of the manuscript and valuable suggestions. I would like to thank especially Dr. I Vaara for the cooperation during model building and photographing. I am indebted to Gunnel Johansson for typing the manuscript with undying enthusiasm. The photographic help rendered by H. Ukkonen is gratefully acknowledged. I would like to thank all my colleagues and my wife Vasantha for the support I received during the preparation of the manuscript. I would also like to acknowledge Profs. R.E. Dickerson, J. Drenth, R. Huber, L.H.Jensen, W.N. Lipscomb, B.W. Matthews and H.C. Watson for permission to reproduce published and unpublished material.

This work was supported by grants from the Faculty of Science, University of Uppsala and the Swedish Natural Science Research Council.

Table 7. Matrix for comparison of crystallographic and chemical sequence data Lactate dehydrogenase

	A	B	C	D	E	F	G	H	I	K	L	M	N	P	Q	R	S	T	V	W	X	Y	Z	
Sequence No.	Ala	Asx	Cys	Asp	Glu	Phe	Gly	His	Ile	Lys	Leu	Met	Asn	Pro	Gen	Arg	Ser	Thr	Val	Trp	unknown	Tyr	Glx	Final grade[a]
44				○													B-C	B	B					C
45		C		C							(B)		C											B
46	(B)						A																	B
47		C		(C)	C				B		C		C		C			B	B				C	B
48					○	B		B				B										A		B
49																	C	B	(B)					B
50	(B)										○						C							A
51																	B	C	C					B-C
52	C																B		○					B-C
53		B		(B)	C			C			B		B		C								C	A-B
54		C		C									C				B	B	(B-C)					B-C
55	C-D	C-D		C-D	C-D		B					○	C-D		C-D								C-D	B-C
56	B				○												C							E
57	B-C			○																				E
58		C	C	C									C				(B)	B	B					B-C
59		B-C		B-C	C		A-B			○○○			B-C		C								C	A
60	C						C																	D-E
61	D-E						(A)																	A
62					(B)							B-C			B					C-D		B	B	B
63					C		A-B					(C)			C								C	B-C
64	B											○					C							E
65		B		(B)							C		B											A-B

66										A		○						C	C			C
67						Ⓒ							C		C	B–C					C	B
68						C		C	○						C						C	C
69		E						Ⓐ														A
70																	Ⓑ	B	B			B
71						B			D			Ⓓ			B						B	C
72						C	Ⓐ		B						C					C	C	A
73			C	B	C							Ⓓ		C			C					C–D
74						C			○				B		C						C	C
75																		C	Ⓒ			C
76	○																B					B
77											○						C	C	C			B–C

Grading: A shape and volume fit well. B all atoms in density but shape of electron density ill defined. C 1–2 atoms outside density, or some excess density. D a few atoms outside ill-defined density, or a significant amount of excess density. E electron density absent but space and environment would not exclude residue. Blank, unlikely or not considered.

[a] Grade given when sequence was known.

Table 8. *Paracoccus denitrificans* cytochrome c_{550} sequence. Tryptic (T) and Chymotryptic (C) peptides are marked.

Acetyl Asn Glu Gly Asp Ala Ala Lys Gly Glu Lys Glu Phe Asn Lys Cys 15

T3 T8ch2 T10ch2

Lys Ala Cys His Met Ile Gln Ala Pro Asp Gly Thr Asp Ile Lys 30

T14

C33b

Gly Gly Lys Thr Gly Pro Asn Leu Tyr Gly Val Val Gly Arg Lys 45

T9ch2 T7,8ch1

C24b

C9 C20b

Ile Ala Ser Glu Glu Gly Phe Lys Tyr Gly Glu Gly Ile Leu Glu 60

T13 T6

C22

C14b C35 C3

Val Ala Glu Lys Asn Pro Asp Leu Thr Trp Thr Glu Ala Asn Leu 75

T11ch1

Ile Glu Tyr Val Thr Asp Pro Lys Pro Leu Val Lys Lys Met Thr 90

T5

C12 C24d C33a

Asp Asp Lys Gly Ala Lys Thr Lys Met Thr Phe Lys Met Gly Lys 105

T4 T9ch1 T10ch3 T12 T11ch3

C20a C24c C33a

C28b C33a

Asn Gln Ala Asp Val Val Ala Phe Leu Ala Gln Asx Asx Pro Asx 120

T1

C8a C1

Ala Gly Glx Gly Glx Ala Ala Gly Ala Gly Ser Asx Ser Glx 134

Amino Acid Sequence Determination by Mass Spectrometry

H. FALTER

I. Introduction

In order to understand the discussion on the application of mass spectrometry to the analysis of the amino acid sequence of proteins, the reader must be familiar with the basic principles of mass spectrometry. The following description of the design and the operation of mass spectrometers is intended to provide sufficient background for this purpose. For more extensive treatments of this subject, the reader is referred to a number of excellent basic texts (42, 43, 45, 72, 175, 235, 297, 377).

II. Instrumentation and Techniques

A mass spectrometer, regardless of the specific design of the instrument, consists of a number of basic units as shown in a block diagram in Figure 1

Fig. 1. Block diagram of a mass spectrometer

The mass analyzer and, in most mass spectrometers, also the ion source, must be maintained under a high vacuum. Typically, in an operating mass spectrometer, the pressure is less than 10^{-6} mm Hg. Excepting some special applications, the sample must be introduced into the ion source in gaseous form. The requirement of a gaseous sample has substantial implications for the analysis of peptides, which are, by their very nature, of relatively low volatility. Once the sample is vaporized into the high vacuum of the ion source the neutral sample molecules are converted to ions by a variety of methods.

A. Ionization

In those instruments currently most widely used for the analysis of peptides, the sample molecules are bombarded by a stream of electrons.

Abbreviations: TFA Trifluoroacetyl; Ac Acetyl; m/e mass to charge ratio; EI Electron Impact; CI Chemical Ionization; FD Field Desorption; FI Field Ionization; MS Mass Spectrometry; GC Gas Chromatography.

The electrons are released from a heated filament and typically accelerated through a potential difference of 70 volts. When electrons of this energy strike organic molecules both positive and negative ions are produced:

$$M + e^- \longrightarrow M^-$$
$$M + e^- \longrightarrow M^+ + 2e^-$$

For the analysis of organic compounds the positively charged ions are of greater importance than the negative ones and most mass spectrometers are constructed to analyze the cationic component. This is especially true when applied to the analysis of peptides. Ions possessing more than one positive charge can be formed by electron impact (EI) but these are generally of low abundance. For most practical purposes the ions analyzed may be considered to possess a single positive charge.

Alternate methods for the production of ions exist. One of these, chemical ionization (CI) is becoming increasingly important. In this method the sample is ionized not by bombardment with electrons but by reaction with other ions. One of the usual sources of such ions is methane gas which is introduced into the mass spectrometer at relatively high pressure in the order of a million times greater than the pressure of the sample. Electron bombardment of this methane-sample mixture produces ions almost exclusively from methane. Under the conditions of pressure which prevail in the ion sources of these instruments, molecular ions of methane collide with other methane molecules to produce ions such as CH_5^+. Protonated methane in turn generates ions from the organic sample molecules by proton transfer:

$$M + CH_5^+ \longrightarrow CH_4 + MH^+$$

The usefulness of chemical ionization lies in the fact that the decomposition of the MH^+ ions proceeds by different routes than the decomposition of the M^+ ions produced by electron impact. Pumps of high capacity are required to maintain the pressure in the mass analyzer below 10^{-6} mm Hg despite the relatively high pressure in the ion source.

Field ionization (FI) and field desorption (FD) are additional methods by which ions may be generated. This method is of importance for the production of ions from molecules of low volatility which decompose when heated to the temperatures required for vaporization. The sample is deposited on a point or edge of two closely spaced electrodes and ionization is achieved by a high electric field gradient (145, 468).

B. Analysis

Once the ions are generated, regardless of the specific method, they are removed from the ion source by positively charged repellor plates which drive them into the mass analyzer. Typically, the ions are then accelerated in an electric field of a potential difference V. The potential energy of a particle of charge e which is placed into an electric field of a potential difference V is equal to eV. This potential energy must be equal to the kinetic energy of the ion once it is fully accelerated. We can therefore write the equation:

$$eV = 1/2mv^2 \quad (1)$$

where m is the mass of the particle and v its velocity.

Several methods exist by which the ions can be analyzed. The most common method utilizes a magnetic field. When a particle of charge e travelling at a velocity v enters a magnetic field H it experiences a centripetal force Hev which is counterbalanced by a centrifugal force mv^2, where r is the radius described by the particle and m its mass. Therefore, one can equate these two forces:

$$Hev = \frac{mv^2}{r} \tag{2}$$

Rearranging this equation and cancelling v reduces the equation to

$$r = \frac{mv}{He} \tag{3}$$

Since the velocity of the ion entering the magnetic field is known from Eq. (1), the term v can be replaced with the other parameters of this equation which leads, after rearrangement, to the expression:

$$\frac{m}{e} = \frac{H^2 r^2}{2V} \tag{4}$$

It is clear from Eq. (4) that singly charged ions of mass m, which are accelerated through a constant potential difference V and a constant magnetic field H will describe a flight path the radius of which depends on the mass of the particle m. Clearly, the same radius is described by a particle which has twice the charge, but also twice the mass. An analysis of the radius r provides, therefore, the mass to charge ratio (m/e) of the particle and not, strictly speaking, its mass. Figure 2 illustrates the trajectories described by a beam of ions which have been accelerated by a constant electric field V as they traverse a magnetic field of constant strength H. The $\frac{m}{e}$ ratios decrease from $\frac{m_1}{e}$ to $\frac{m_4}{e}$.

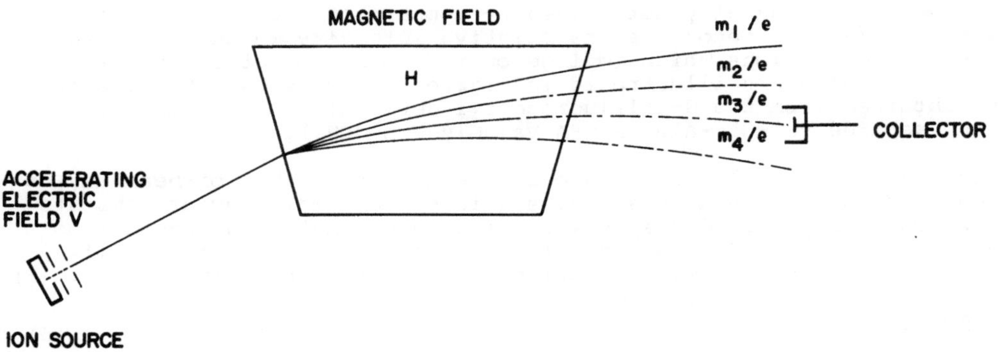

Fig. 2. Trajectories of ions in a magnetic field

It is impractical in most mass spectrometers to measure the radius described by various ions under the influence of a constant magnetic field and a constant accelerating voltage. Such an analysis would require that the detector be moved. It is more practical to have a fixed detector and to vary the magnetic field while keeping the accelerating voltage constant. At different magnetic field strengths, ions of various m/e will impinge on the collector. In recording a mass spectrum it is customary to increase the magnetic field strength at a uniform rate and to record the response of the collector by means of a recorder. This process is called scanning. The recorder output contains essentially only two basic pieces of information. One is the m/e of the

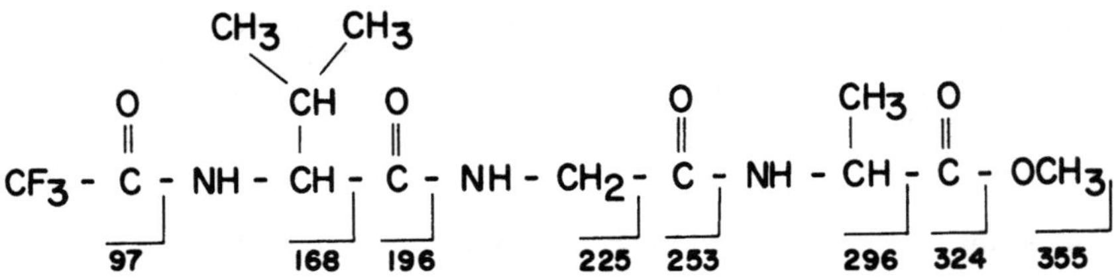

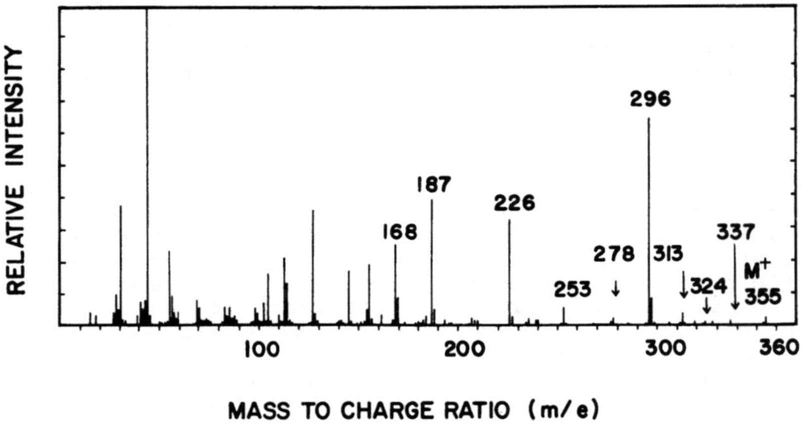

Fig. 3. Mass spectrum of N-TFA-Val-Gly-Ala-OCH_3 [adapted from Ref. (419)]

ions, which is generally determined by comparison with ions of known values of m/e. The second is the relative intensity which is related to the number of ions which impinge on the collector at a given time. These principles are illustrated in Figure 3 which shows the mass spectrum obtained from the N-trifluoroacetyl derivative of the methyl ester of the peptide Val-Gly-Ala (N-TFA-Val-Gly-Ala-OCH_3).

To increase the resolution of a mass spectrometer, the ion beam may be allowed to pass through an electrostatic sector. This sector consists of two uniformly curved plates which are maintained at a constant potential difference. Ions which enter the electrostatic sector halfway between the two plates will describe a circular path defined by Eq. (5)

$$\frac{mv^2}{r} = eV^1 \tag{5}$$

where m is the mass, e the charge, v the velocity of the particle and r the radius of the circle. The term V^1 represents the potential difference between the two plates. From Eq. (1) it can be seen that the velocity of the particle v is given by the expression:

$$v = \left[\frac{2eV}{m}\right]^{1/2} \tag{6}$$

where e and m have the usual meaning and V is the potential through which the ion has been accelerated. Substituting the expression for velocity of Eq. (6) into Eq. (5) and cancelling the appropriate terms leads to an expression which describes the radius of the path of an ion in the electrostatic sector.

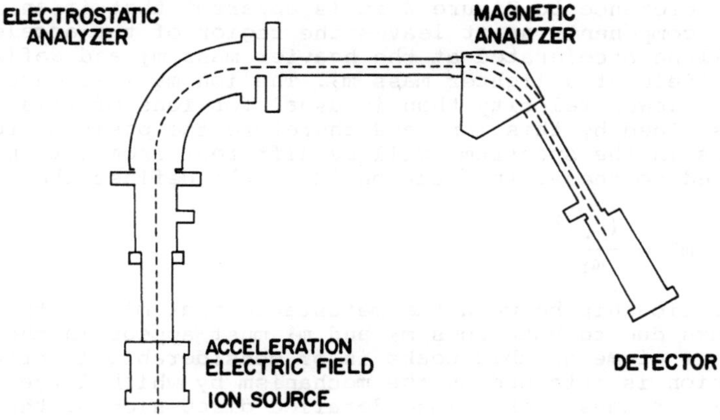

Fig. 4. Double focusing mass spectrometer (Nier-Johnson)

$$r = \frac{2V}{V^1} \tag{7}$$

By placing entry and exit slits at the beginning and end of the electrostatic sector, a beam of ions of narrowly defined energy is available for separation in the magnetic sector.

High resolution instruments typically consist of an electrostatic sector followed by a magnetic sector. Such instruments are called double-focusing instruments. Figure 4 shows in diagrammatic form a double-focusing mass spectrometer of the Nier-Johnson geometry.

A variety of mass spectrometers with mass analyzers which operate on principles different from those discussed above are commercially available. Among these, the quadrupole and the time of flight mass spectrometers are quite common. A discussion of the operational principles of these and others may be found in reference (377).

In most mass spectrometers the detectors are metal electrodes which produce an electrical signal which is amplified by appropriate means. In some double-focusing instruments a photographic plate is used to detect the ions. The degree of darkening produced by ion impact can be translated into relative intensities.

The detector output may be recorded by means of conventional recorders, oscilloscopes or magnetic tapes. In many systems the output from the detector is interphased with a computer, which may process the data as desired.

In addition to the peaks shown in Figure 3 which generally occur at or near unit m/e, mass spectra also contain peaks of entirely different appearance. These additional peaks generally are of low intensity, usually occur in positions which do not correspond to a unit mass to charge ratio, and have broad triangular shapes. These are called metastable peaks and are produced by ions which have life times sufficiently long so that they do not decompose in the ion source of the instrument, but instead in the analyzer section.

By reference to Figure 2 it is apparent that if an ion decomposes into two components as it leaves the region of the accelerating field it will be accelerated at the heavier mass m_1 and deflected in the magnet ic field at a lighter mass m_2. The ion m_2 will enter the magnetic fiel at a lower velocity than is usual for ions of this mass. The radius described by this ion, and therefore the position relative to other ions in the spectrum, will be different from both m_1 and m_2 but is related to these. An equation [Eq. (8)] defines the

$$m^* = \frac{(m_2)^2}{m_1} \tag{8}$$

relationship between the metastable peak m* and the ions m_2 and m_1. Peaks due to both ions m_2 and m_1 must appear in the spectrum. The analysis of metastable peaks is of considerable importance in that information is obtained on the mechanism by which larger ions decompose to smaller ones. For a more detailed discussion of the nature of metastable transitions the reader may refer to a number of recent publication (193, 215, 256).

C. Resolution

The term "resolution" expresses the ability of the mass spectrometer to separate two peaks in a spectrum. Resolution is defined as $\frac{m}{\Delta m}$, wher m is the mass of one of the peaks and Δm is the separation in mass units between the two peaks. Unfortunately there is no general agreement with respect to the criteria for separation. Frequently used definitions rely on the height of the valley between two peaks of equal intensity. Arbitrarily, a peak may be considered separated when the valley between two peaks is 1, 2, 10 or even 50%.

Low resolution instruments typically operate with a resolution in the order of 1000 which is sufficient to provide integral mass resolution for most peptides. High resolution instruments typically operate with a resolution of 10,000 or higher. With high resolution instruments it is possible to identify the molecular composition of a given ion. The large amount of data obtained with a high resolution instrument virtually demands that it be processed by computer. Table 1 shows a typical

Table 1. Partial list of the computer printout of the high resolution mass spectrum of α, ε-diaminopimelic acid

Calculated mass	Error	C_{12}	C_{13}	H	N	O	Measured mass
172.0847	-.88	7	0	12	2	3	172.0838
155.0820	-2.07	7	0	11	2	2	155.0799
155.0775	2.38	6	1	10	2	2	
154.0742	.00	7	0	10	2	2	154.0742
152.0585	.15	7	0	8	2	2	152.0587
129.0789	-1.86	6	0	11	1	2	129.0808
128.0711	2.83	6	0	10	1	2	128.0739

Abstracted from H. Falter (276).

computer printout of a portion of a high resolution mass spectrum obtained from α,ε-diaminopimelic acid.

D. Sample Introduction

Amino acids and peptides are of relatively low volatility even if they are suitably derivatized. Solid materials can be introduced into the mass spectrometer with a probe to a position only a few mm from the electron beam used to ionize the sample. The probe is then heated and the sample is ionized as soon as it evaporates. An alternate and increasingly more important method for the introduction of the sample into the mass spectrometer utilizes the direct combination of a gas chromatograph (GC) to the mass spectrometer (MS). One of the prerequisites to such a coupling is the reduction of the atmospheric pressure of the GC effluent to the operating pressure of the mass spectrometer with only minimal loss in the partial pressure of the component of interest. In practice this requires that the GC carrier gas be removed preferentially. A number of systems are now available which meet this requirement (380). One of the more frequently used coupling systems is the Watson-Bieman separator, which depends for its function on the difference in the diffusion rate across the pores of a fritted glass tube between the carrier gas helium and the sample (461). The direct coupling of gas chromatography and mass spectrometry, especially when interphased with a computer, opens the way for the rapid analysis of very complex mixtures of organic materials. The hydrolysates generated by the partial degradation of large peptides or proteins represent such complex mixtures.

III. Mass Spectrometry of Peptides

A. Derivatization of Peptides

A majority of peptides which one might obtain from the hydrolysis of a protein are not of sufficient volatility to be analyzed directly by mass spectrometry. Underivatized peptides, even relatively small ones, are prone to thermal degradation if they are heated to the temperatures required to vaporize a sufficient amount of the sample to obtain a usable spectrum (221).

The reasons for the low volatility of peptides are to be found in their polar nature. Peptides possess at least one positive charge at the N-terminus and one negative charge at the C-terminus (see Fig. 5). In addition, depending on the amino acid composition of the peptide, additional ionic groups may be contributed by the various side chains. In general it is necessary to derivatize peptides to convert the ionic groups into nonionic ones in order to increase the vapor pressure of the peptide sample to the point where mass spectra can be obtained without extensive thermal degradation.

The derivatization of the C-terminal carboxyl, and possible side chain carboxyl groups, is most easily accomplished by esterification. Esterification can be carried out on the microscale required for peptide work (146, 392, 400). Special care is indicated with peptides containing glutamine and asparagine residues to avoid solvolysis of the side chain amides (477). A variety of esters, such as methyl, ethyl and t-butyl esters have been examined for their suitability to the mass spectrometry of amino acids and peptides (276, 404, 419). Methyl or ethyl esters of peptides are satisfactory derivatives for mass spectrometry and are widely used.

A great deal of attention has been devoted over the years to the natur of the N-blocking group. Such attention is justified because the natur of the N-blocking group not only affects the volatility, but also the fragmentation pattern of the peptide (106, 174, 176, 331, 336, 356, 392). Some of the N-blocking groups, which were investigated very earl such as the acetyl (Ac) (189) and trifluoroacetyl (TFA) groups (15, 41 465, 466) are still widely used because of the relative ease with whic these derivatives can be prepared.

B. Determination of Amino Acid Sequence from the Mass Spectrum

1. *N-Acyl Peptide Esters*

a) N-terminal Sequence Peaks. The first peptide derivatives to be studied by mass spectrometry were trifluoroacetyl peptide esters (15, 419). These are readily prepared from the peptide in two steps as outlined in Figure 5. The resulting derivative is volatile enough for analysis by mass spectrometry.

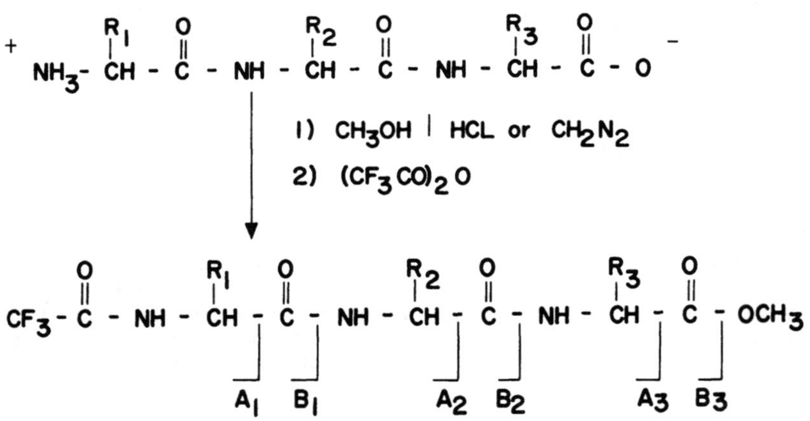

Fig. 5. Preparation of a volatile peptide derivative (N-TFA peptide methyl ester)

Electron bombardment of this derivative in the ion source of a mass spectrometer will generate a positive ion (I) by removal of an electro The spectrum peak corresponding to ion (I) is called the molecular ion peak M^+. The M^+ peak is usually of low intensity because ion (I) will decompose further.

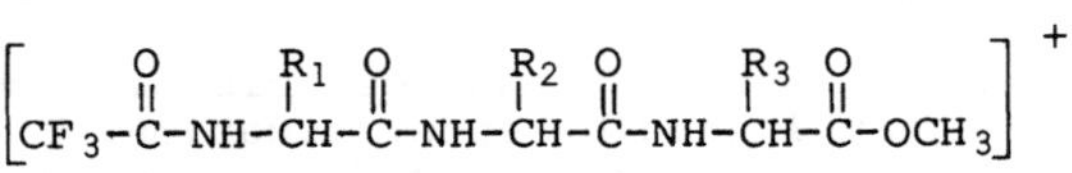

(I)

Cleavage at position B_3 (see Fig. 5) with retention of the positive charge on the N-terminal fragment leads to an ion (II) whose mass is M^+-31. Such fragments are frequently referred to as amino acid frag-

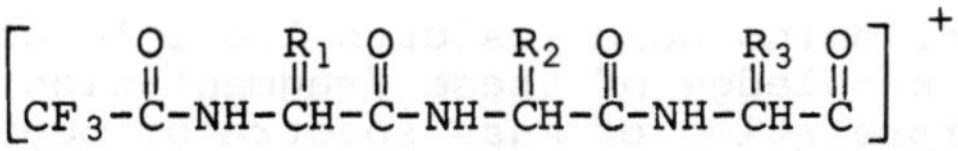

(II)

ments. Ion (II) may in turn decompose, with the loss of CO to ion (III). Ions of this type are called aldimine fragments. The sequential fragmentation has been confirmed in a number of cases by the presence of

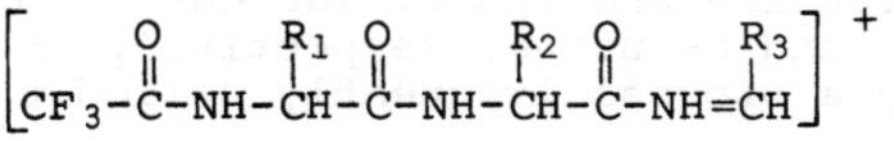

(III)

the appropriate metastable peaks. The further loss of a neutral fragment of the structure HN=CH-R_3 through a cleavage at B_2 leads to a new amino acid fragment, the mass of which corresponds to the residue weight of all but the C-terminal amino acid (IV).

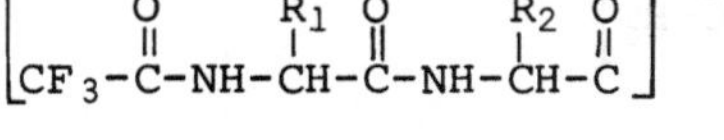

(IV)

The fragmentation process may be initiated in various positions in the peptide resulting in the representation in the spectrum of many, although certainly not always all, of the fragments due to cleavage at positions A and B. Peaks at m/e values of 168, 225 and 296 in Figure 3 correspond to cleavage at positions A_1, A_2, and A_3, respectively; those at 196, 253 and 324 correspond to the B series of fragments. In general, if the amino acid composition of the peptide is known, then the amino acid sequence can be deduced from the spectrum by a consideration of the combinations of peaks which would arise from cleavage at positions A and B.

Even a cursory glance at Figure 3 indicates that the peaks due to fragmentation at positions A and B are not the only ones found in the spectrum. Frequently one finds peaks which are ostensibly due to cleavage at positions A but with a mass which is one atomic mass unit higher than that calculated for the fragment. Weygand et al. (465) have demonstrated that these arise from the removal of a hydrogen atom from the nitrogen of the peptide bond immediately following the fragment in question. The abstraction of the hydrogen is postulated to occur in a cyclic transition state with retention of the charge on the N-terminal fragment. The peak at m/e 169 in Figure 3 can be attributed to cleavage of this type.

b) Additional Peaks. The presence of peaks 18 mass units below that of the M^+ peak is indicative of dehydration which is common in peptides requiring heating to relatively high temperatures in order to obtain a spectrum. Dehydration involves the removal of a carbonyl oxygen from a peptide link along with two hydrogens from adjacent atoms.

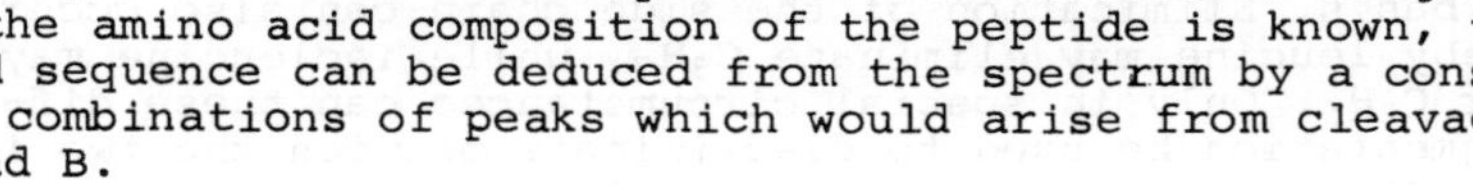

(V) (VI)

Since dehydration involves internal peptide bonds, a second series of peaks, due to cleavage at positions A and B, is expected at 18 mass units lower than their usual mass. The small peaks at m/e 337 and 278 in Figure 3 can be attributed to dehydration. Dehydration is favored by the presence in the peptide of residues of glycine, aromatic and heterocyclic amino acids (406).

Fragmentation of the side chains of amino acid residues leads to additional peaks in the spectrum. A knowledge of these fragmentation patterns is important to the interpretation of mass spectra of peptide since the peaks produced by these are frequently quite prominent. It should be noted, however, that the fragmentation of side chains is strongly influenced by the nature of the N-terminal blocking group (10

The aliphatic amino acids, valine, leucine and isoleucine can eliminat the side chain as propene, isobutene and n-butene, respectively. Elimination of the side chain proceeds by a process frequently called the McLafferty rearrangement (296).

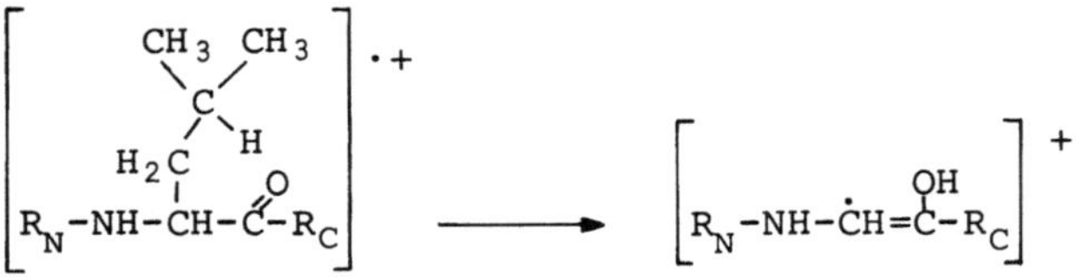

(VII) (VIII)

Figure 3 shows a peak at m/e 313 which can be attributed to the loss of propene from the M^+ ion.

Leucine and isoleucine residues yield identical sequence peaks due to cleavage at positions A and B and can for that reason not be distinguished on that basis. Elimination of the side chain can also occur as a radical, whereby leucine may eliminate C_3H_7, while isoleucine may eliminate CH_3 or C_2H_5. Only in special circumstances can these differences in fragmentation be used to discriminate between the two isomers (465, 466). Recently, Levsen et al. (256) have described a method by which leucine and isoleucine residues can be distinguished on the basis of their metastable and collisional activation spectra. Falter and Waern (147) have reported a study in which synthetic peptides containing leucine and isoleucine were derivatized with N-TFA-perdeutero-Leu as the N-terminal blocking group. The resulting spectra showed peaks due to the elimination of the leucine or isoleucine side chains of the original peptide (M^+ - 56) and peaks due to the loss of the deuterated side chain (M^+ - 64). A wide variation in the relative intensities of these two peaks were reported, but for a given position in a peptide chain the side chain of leucine was always more readily eliminated than the side chain of isoleucine. The nature of the N-terminal blocking group is important with respect to the elimination of side chains of leucine and isoleucine. Day et al. (106) reported that the ion (M^+ - 56) which results from elimination of C_4H_8 from the side chain of isoleucine could not be observed in the DNS [5-(N,N-dimethylamino)naphthalenesulfonyl] methyl ester derivative of the model peptide Val-Ile-Ala, while many other derivatives showed relatively intense peaks due to this fragment.

Side chains containing functional groups in the β-position such as serine, threonine and cysteine may eliminate these groups as H_2O or H_2S.

X=OH or SH
R=H or CH_3

(IX) (X)

In the case of serine and threonine the side chain may be lost in a process which involves the migration of a hydrogen atom.

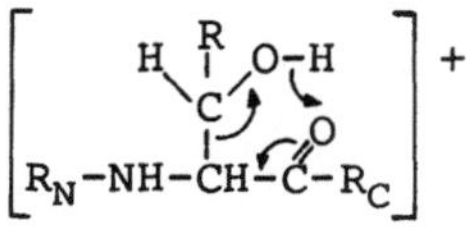

(XI)

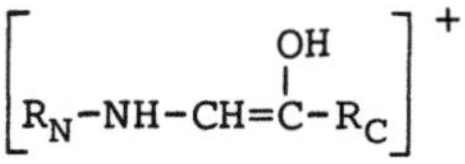

(XII)

In proline-containing peptides, the aldimine fragment due to cleavage at position A may decompose to yield ions (XIV) and (XV) in low yields.

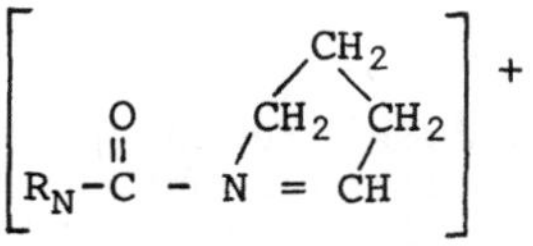

(XIII)

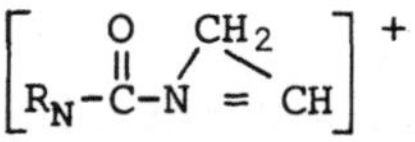

(XIV)

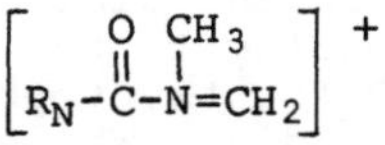

(XV)

Peptides containing the esterified residues of glutamic and aspartic acid or their amides, i.e. asparagine and glutamine, are of special interest because in the course of conventional acid hydrolysis the side chain amides are hydrolyzed requiring additional steps to establish the identity of these residues. The side chain methyl esters of glutamic and aspartic acid eliminate methanol under electron impact, while glutamine and asparagine lose ammonia usually in a thermal process. Both processes lead to an intermediate ion (XVI) of identical mass for the ester and the amide of the two pairs of residues.

n = 0 or 1

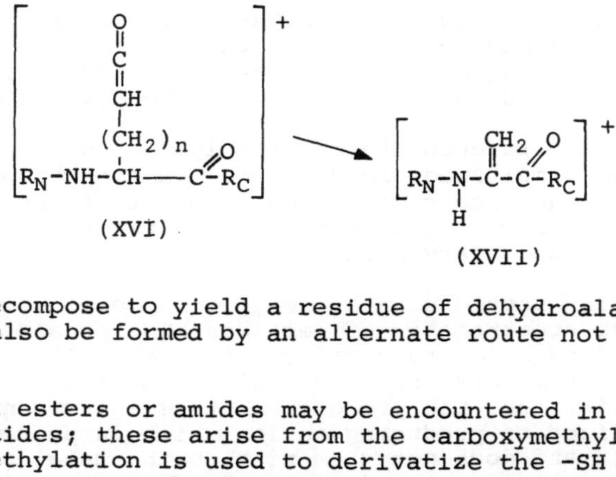

(XVI) (XVII)

Ion (XVI) can decompose to yield a residue of dehydroalanine (XVII). Ion (XVII) can also be formed by an alternate route not involving ion (XVI) (406).

Other side chain esters or amides may be encountered in the mass spectrometry of peptides; these arise from the carboxymethylation of cysteine. Carboxymethylation is used to derivatize the -SH groups of cysteine in order to prevent the formation of disulfide bridges; this is frequently carried out prior to the enzymatic degradation of proteins. Carboxymethylcysteine residues (XVIII) may eliminate CH_2-COX to form ion (XIX) as shown or may eliminate $HSCH_2$-COX which leads to an ion containing dehydroalanine (XVII).

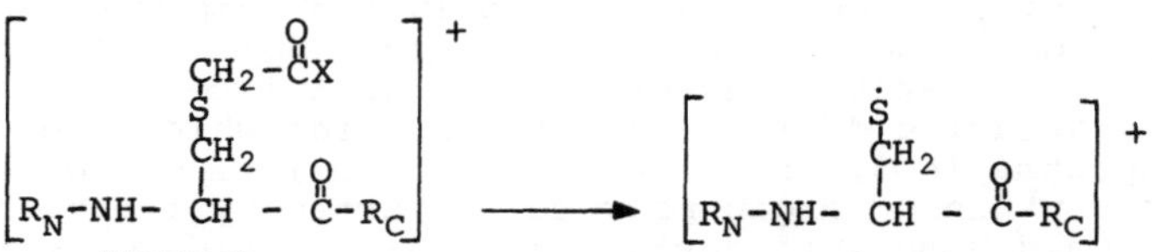

(XVIII) (XIX)

X = OCH_3 or NH_2

Methionine-containing peptides show strong peaks due to loss of the entire side chain (XX) → (XXI) and peaks which result from the cleavage of the carbon-carbon bond connecting the β- and γ-carbons and the γ-carbon-sulfur bond.

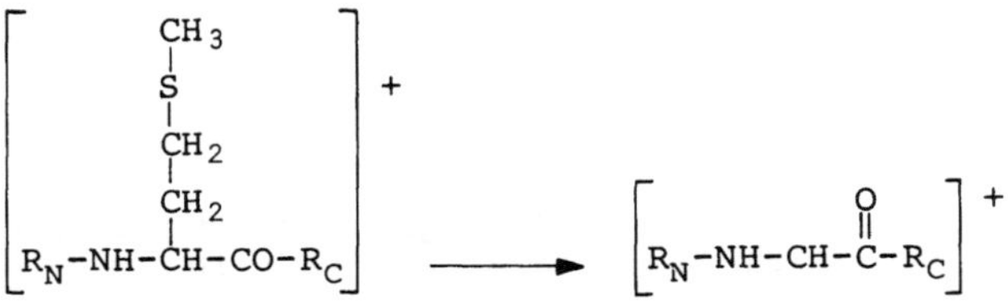

(XX) (XXI)

Peptides containing aromatic and heterocyclic residues display peaks due to the ionized side chain. Evidently during electron bombardment the charge may be localized in the side chain which is then eliminated as $ArCH_2^+$. The ions corresponding to the various side chains have characteristic m/e values of 130 for tryptophan, 107 for tyrosine, 91 for phenylalanine and 81 for histidine. While these ions are very intense and serve to identify the presence of these amino acids they are, of course, not useful for sequence determination.

The side chain may be eliminated as a radical of the type $ArCH_2^{\cdot}$ from peptides containing phenylalanine and histidine which leads to peaks at M^+-91 and M^+-81, respectively. Elimination of the side chain occurs with the transfer of a hydrogen to the remaining positive fragment from peptides containing tyrosine and tryptophan which generates intense peaks at M^+-106 for the former and M^+-129 for the latter.

Histidine-containing peptides are characterized by the occurrence of a series of peaks 14 mass units higher than the calculated molecular ion and the various fragment peaks. These peaks arise from the partial methylation of the imidazole ring of histidine (342, 400, 404).

No special problems are encountered with lysine as long as the ε-amino group is acylated simultaneously with the α-amino group of the N-terminal amino acid. The side chain shows cleavage of the C-C and C-N bonds accompanied frequently by hydrogen ion migration characteristic for the pattern of secondary amides (162).

Arginine-containing peptides have always occupied a somewhat special place in the mass spectrometry of peptides. The presence of the very basic unprotected guanido group reduces the volatility of the peptide to such an extent that extensive thermal decomposition occurs in the process of vaporization. The presence of three nitrogens in the side chain has the effect of producing an ion which is predominantly ionize in the side chain and does not give rise to the usual peaks due to fragmentation at positions A and B. To overcome these difficulties several methods have been suggested which lead to interpretable mass spectra of peptides containing arginine. Shemyakin et al. (407) suggested the conversion of the arginine residues to ornithine by treatment with aqueous hydrazine. The resulting peptides, now containing ornithine instead of arginine, could be handled in the same way as lysine-containing peptides. The condensation of arginine-containing peptides with either 1,1,3,3-tetralkoxypropane or acetylacetone leads to stable, relatively volatile derivatives of ornithine which yield interpretable mass spectra (405, 451). Day et al. (107) reported recently that peptides containing arginine were converted by treatment with 2-pyridinecarboxyaldehyde into derivatives which could be separat by gas chromatography. The derivatives retained sequence information regardless of the position of arginine in the chain.

c) C-terminal Sequence Peaks. In N-acylated peptides, depending on the nature of the blocking group, the charge is usually retained by the N-terminal fragment. Peaks due to the elimination of a portion of the N-terminus of the peptide and retention of the charge on the C-terminal fragment are usually of low intensity in mass spectra generated by electron impact. C-terminal fragments arise mainly in peptides containing aromatic or heterocyclic amino acids. On the other hand, C-terminal fragments are common in Schiff base derivatives of peptides (174, 176, 336) and are an important factor in interpreting such spectra. In these derivatives "internal fragments" possessing neither the N- nor the C-terminal portion of the peptide may be crucial to the assignment of the sequence. The occurrence of certain single peaks in a spectrum can sometimes be attributed to a particular amino acid sequence (113).

2. Permethylated Peptide Derivatives

While N-acylated peptide esters give rise to interpretable mass spectra, it is nevertheless true that the volatility of these derivatives is still relatively low. Mass spectra of such derivatives, which contain in excess of seven or eight amino acid residues, cannot usually be obtained even if the sample is introduced directly into the ion source with a solid insertion probe. Van Heijenoort et al. (446) suggested that one factor in the low volatility of these derivatives might be the hydrogen bonding due to peptide linkages. Each peptide link contributes a polar -CO-NH- group at regular intervals which are ideally suited to form hydrogen bonded lattices. Once this factor was recognized it was suggested that methylation of the peptide nitrogen would lead to derivatives of higher volatility, which should yield suitable mass spectra. The correctness of this reasoning was demonstrated by Das et al. (103). The heptapeptide derivative (XXII) did not give a mass spectrum owing to its low volatility. Permethylation, by the procedure of Kuhn et al. (240), using methyl iodide in dimethylformamide with silver oxide as a catalyst, led to a derivative (XXIII) which gave a satisfactory mass spectrum (Fig. 6).

The permethylated derivative displayed a molecular ion peak at m/e 1082 corresponding to an expected increase in molecular weight of 112 as a result of the methylation of seven nitrogens and one oxygen. The mass spectrum provided clear evidence that the methylation had proceeded to completion. In addition it became apparent that permethylation had simplified the mass spectrum. Prominent peaks were observed at m/e 43, 114, 275, 436, 627 and 756 which correspond to cleavage at the peptide bonds and retention of the charge on the N-terminal fragment. Peaks due to cleavage at other positions were substantially reduced. In part, the simplification of such spectra may be attributed to the lower temperature which is required to vaporize the permethylated derivatives which inhibits thermal degradation. Improvements in the original silver oxide catalyzed permethylation reaction were soon reported. Vilkas and Lederer (452) demonstrated that the method described by Hakamori (173) for the methylation of glycolipids is applicable to peptide derivatives. The method employs NaH in dimethylsulfoxide together with CH_3I to effect methylation. According to Thomas (431), this method is also suitable for peptides containing aspartic acid, glutamic acid and tryptophan for which the silver oxide catalyzed reaction frequently leads to undesirable by-products (7, 431). More recently Morris et al. (304, 307) have reported a short permethylation procedure by which peptides containing the problematical amino acids histidine, arginine,

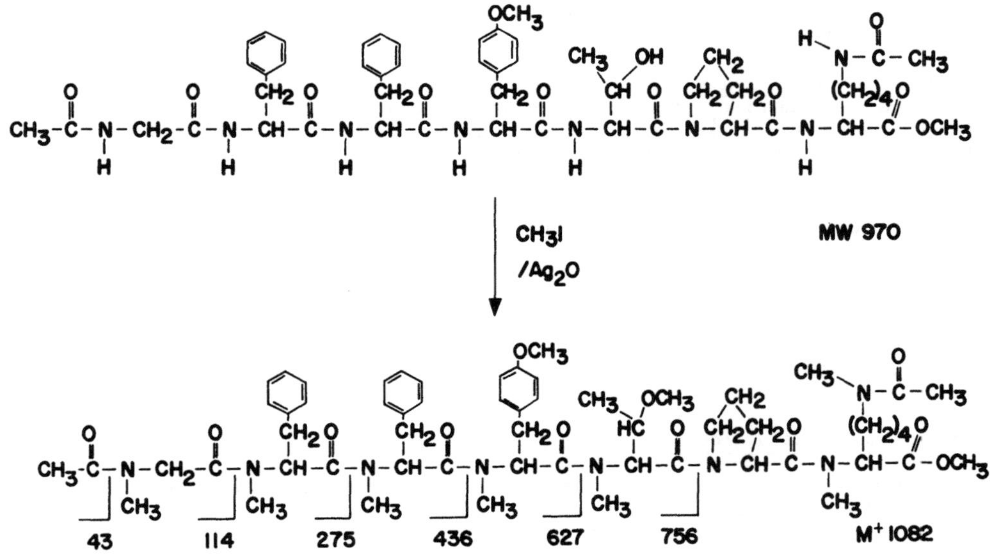

Fig. 6. Permethylation of a heptapeptide derivative

methionine and carboxymethylcysteine could be derivatized. This procedure utilizes an excess of base and CH_3I but a very short reaction time in the order of 60 s. Under these conditions the extensive formation of involatile sulfonium "salts" in the case of peptides containing methionine and cysteine, and quaternary ammonium salts in the case of histidine, is avoided. The method has the advantage over previously reported (for details see references quoted in (304)) improvements in the permethylation procedure, that an exact knowledge of the amount of peptide present in order to balance the amount of base, CH_3I and peptide is no longer required. The exact quantity of peptide to be derivatized is frequently not known for peptides isolated from proteins by various methods of hydrolysis. Carboxymethylcysteine gave rise to dehydroalanine in the two cases studied. Arginine-containing peptides can be handled either by condensation of the guanidine side chain with acetylacetone to pyrimidyl ornithine or by hydrazinolysis of arginine to ornithine followed by acetylation of the new amino group. Detailed procedures were reported for both of these steps which yielded satisfactory results in the hands of these researchers. Mass spectra which could be interpreted with respect to amino acid sequence were obtained from peptides treated by either procedure.

Permethylation has been widely used since its introduction for the analysis of synthetic peptides, peptides derived from proteins and particularly for the analysis of peptide antibiotics. This subject has been reviewed by Das and Lederer (104).

The size limitations of peptides which can be analyzed after permethylation became apparent relatively early after the introduction of the permethylation technique. A few of the examples reported by Thomas et al. (433) may serve as an illustration: Ac-Leu-Ala-(TFA)Lys-Val-Ala-Tyr-Val-Tyr-(TFA)Lys-Pro yielded after permethylation a mass spectrum which showed the expected molecular ion peak at m/e 1580 and sequence peaks which allowed the deduction of the entire sequence.

The naturally occurring antibiotic stendomycin, a pentadecapeptide, gave a mass spectrum after permethylation which allowed the deduction of the sequence of the first 10 amino acids without prior knowledge of the sequence (432).

A synthetic tetradecapeptide of the sequence Ac-Gly-Ile-Gly-Ala-Val-Leu-(Ac)Lys-Val-Leu-Thr-Thr-Gly-Leu-Pro yielded a mass spectrum which showed sequence peaks up to and including the first 11 amino acid residues. The pentadecapeptide derivative gramacidin and several analogs gave volatile derivatives and showed sequence peaks for the first twelve residues (104, 433).

It was apparent that all these derivatives had appreciable volatility after permethylation, but that the limitation now arose from the inability to detect the small number of ions which corresponded to the fragments higher than the tenth or twelfth residue. It is well known that the intensity of successive sequence ions decreases and at some point these ions may represent such a small part of the total ion current that they are no longer detectable above the background noise.

3. Polyamino Alcohols

The volatility of peptide derivatives can also be increased by the reduction of the peptide links to form polyamino alcohols (45). Biemann et al. (44) recognized the value of these peptide derivatives as far back as 1959. Their preparation according to the most recent reports (318-321) typically involves four steps: (a) esterification with methanolic HCl, (b) acylation with either methyl trifluoroacetate, or the anhydrides of pentafluoropropionic or heptafluorobutyric acid, (c) reduction with $LiAlD_4$, and finally, (d) selective o-trimethylsilylation. This sequence of reactions will convert peptide (XXIV) into derivative (XXV) or the appropriate higher fluorine analogs.

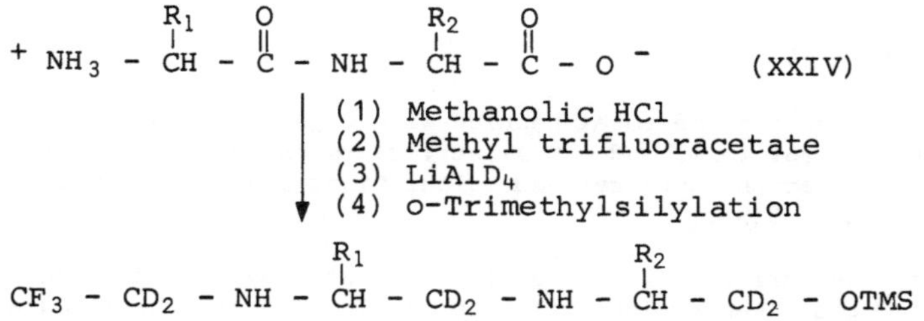

(XXIV)

(XXV)

Derivatives such as (XXV) have a number of attractive properties:

a) They are quite volatile which permits their gas chromatographic separation, as long as the original peptides are relatively small; Derivatives of peptides containing all common amino acids, including glutamine, asparagine, histidine, tryptophan and arginine are amenable to gas chromatography; Only two types of peptides were found to give more than one product following derivatization.

Peptides with N-terminal proline showed, in addition to the expected derivative (XXVI), a product in which proline had no N-terminal blocking group (XXVII). In fact, the latter compound constituted the major product and was generally used for the interpretation of the results; the normally expected product (XXVI) was used to confirm the assignment.

Peptides containing arginine gave two products as a result of the derivatization procedure both of which are formed in low yield. Both products are related to orni-

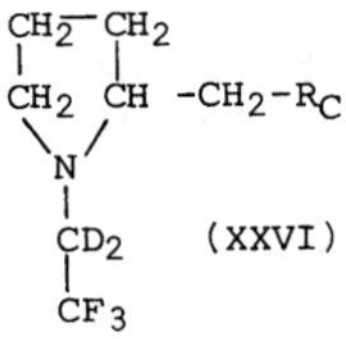

(XXVI)

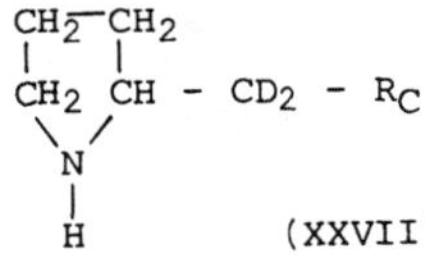

(XXVII)

thine; the major one is the ω-trideuteromethyl (XXVIII) and the minor one, the ω-trifluorodideuteroethyl (XXIX) derivative of ornithine.

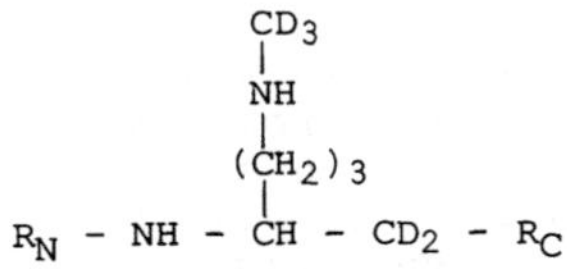

(XXVIII)

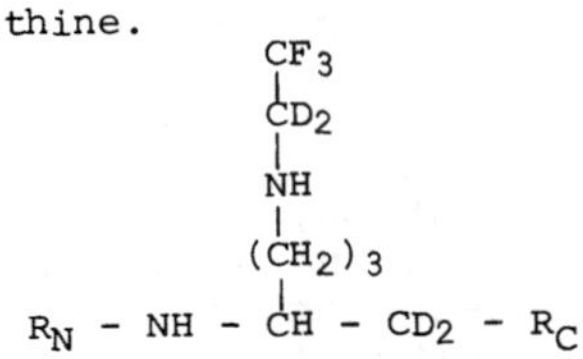

(XXIX)

Pretreatment of arginine-containing peptides to form ornithine would be expected to generate a single product in high yield (407).

b) The gas chromatographic behavior of the polyamino alcohol derivatives is predictable; the retention indices of a particular peptide derivative can be related to the amino acid composition of the original peptide and can therefore be used as a parameter for the identification of the peptide (319, 323). This method of analysis can be applied to other peptide derivatives as well (238).

The retention index of a given o-trimethylsilylated trifluorodideuteroethyl polyamino alcohol can be calculated from empirically determined retention indices of each amino acid residue and a constant term which represents the N- and C-terminal groups. The retention indices for all common amino acids and for some derivatives were reported. From a knowledge of these the retention index of the peptide derivative can be predicted from Eq. (9):

$$I_1(CF_3\text{-}D_2\text{-}) = 630 + \Delta_1 + \Delta_2 + \ldots \Delta_n \qquad (9)$$

where I_1 is the retention index for the total peptide with CF_3D_2- as the N-terminal blocking group, 630 is a constant term, and Δ_1, Δ_2 etc. represent the indices for the various constituent amino acids. More refined calculations and the limitations of the technique are discussed (319).

The retention indices for the various derivatives are generated by co-injecting into the gas chromatograph selected alkanes along with the sample. The gas chromatographic effluent is monitored by a mass spectrometer which is directly coupled to this instrument, which in turn is controlled by a computer. Mass spectra are obtained over the entire range of the temperature programmed GC run and in intervals of 4.7 s. The computer identifies the reference compounds by their mass spectral characteristics and uses their positions to assign retention indices for the entire gas chromatogram (322).

c) The mass spectra of derivatives of the type (XXV) are relatively simple and can be readily interpreted in terms of the sequence of the original peptide. Figure 7 shows the structure and the mass spectrum of the derivative of a peptide with the sequence Ser-Ala-Ala-Ala-Lys. It is clearly evident that cleavage occurs primarily through rupture of the carbon-carbon bonds of the ethylenediamine backbone. The charge may be retained by either of the two fragments which gives rise to two series of ions, one representing the N-terminal fragments designated as A, the other representing the C-terminal fragments Z. Both series are represented, even though the N-terminal series is more intense and can be used to deduce the sequence.

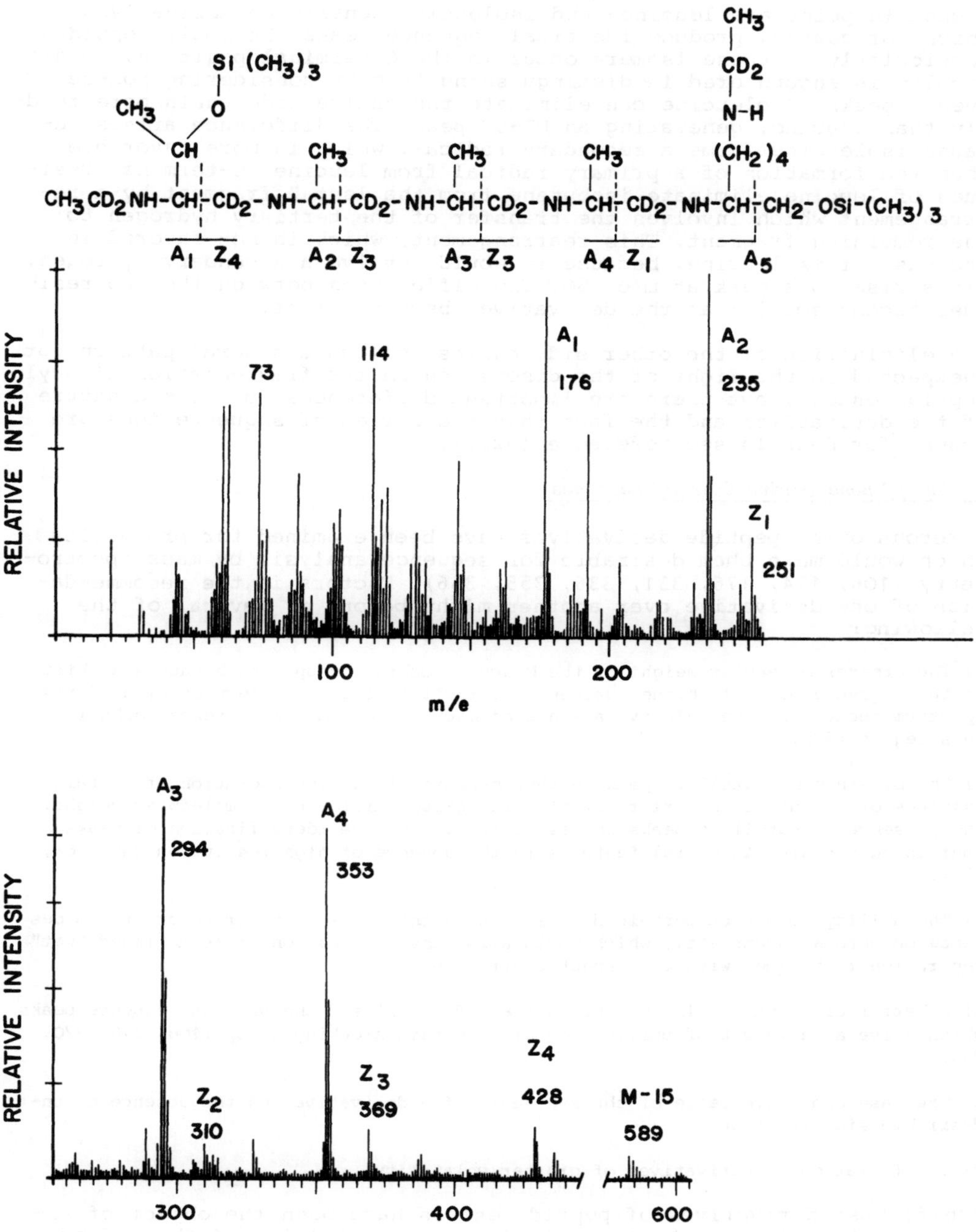

Fig. 7. Mass spectrum of dideuterioethyl polyamino alcohol o-trimethylsilyl derivative of peptide Thr-Ala-Ala-Ala-Lys [adapted from Ref. (317)]

Additional peaks produced by fragmentation of the side chains of various amino acid residues do not interfere with the interpretation of the spectra; in fact, they sometimes aid in the analysis of a spectrum (320).

A case in point are leucine- and isoleucine-containing derivatives which, of course, produce identical sequence peaks. In small peptides, particularly if these isomers occur in the C-terminal position, no difficulty is encountered in distinguishing them by considering non-sequence peaks. Isoleucine can eliminate the entire side chain more readily than leucine, generating an M^+-57 peak. The difference arises because isoleucine forms a secondary radical, which is more favorable than the formation of a primary radical from leucine. C-terminal residues of leucine eliminate isobutene from the last Z fragment by a rearrangement which involves the transfer of the tertiary hydrogen to the remaining fragment. This rearrangement, which is not favored in the case of isoleucine, because it would involve a secondary hydrogen, gives rise to a peak at m/e 150. The differences between the two residues become smaller as the derivatives become larger.

The elimination of the other side chains follows a general pattern not unexpected in the light of the discussion on the fragmentation of acyl peptide esters, but there are important differences due to the nature of the derivatives and the fact that two series of sequence ions are found [for details see reference (320)].

4. Schiff Base Derivatives of Peptides

Numerous other peptide derivatives have been examined for properties which would make them desirable for sequence analysis by mass spectrometry (106, 174, 176, 331, 336, 355, 356). Factors in the recommendation of one derivative over another might be one or several of the following:

a) The greater molecular weight of the N-acyl blocking group, which causes a shift in the sequence peaks to higher values thus facilitating the interpretation of the spectrum because of the relative absence of background and non-sequence peaks in this region (68).

b) The presence of satellite peaks either because of the incorporation of certain isotopes or the use of a mixture of blocking groups differing in molecular weight. The presence of satellite peaks serves to facilitate the identification of peaks containing certain structural features in the absence of high resolution data (68, 152).

c) The ability to detect certain derivatives on the basis of their color or fluroescence on chromatograms etc., which would allow the combination of established "wet" separation techniques with mass spectrometry (162).

d) A better distribution in the intensities of the M^+ and the various sequence peaks which arise as a result of utilization of a certain blocking group (106, 176, 320, 356).

e) The ease and convenience of the synthesis of a derivative and the absence of undesirable side reactions.

f) The formation of derivatives of greater volatility.

Schiff base derivatives of peptide esters have been the object of extensive study in recent years. By judicious selection of the aldehyde used for derivatization of the peptide ester it is possible to manipulate virtually all of the parameters which make desirable blocking groups. Schiff base derivatives of peptide esters are formed conveniently by heating equivalent amounts of aldehyde and peptide ester in glacial acetic acid for a few minutes (106). They are of considerable volatility, as indicated by the fact that a good mass spectrum, which included an M^+ peak, could be obtained for the dimethylaminonaphthylidine derivative of the methyl ester of hexaglycine (336) and the more recent demonstration that Schiff base derivatives of peptide esters are amenable to separation by gas chromatography. Aromatic aldehydes, such as N,N-dimethylaminobenzaldehyde, can serve as dual-purpose reagents for both the detection of amino acids and peptides on chromatograms and as N-terminal blocking groups for mass spectrometry. The possession of these two properties allows the combination of wet separation techniques, such as paper chromatography, with mass spectrometry (146).

Mass spectra of Schiff base derivatives of peptides are somewhat more complicated than those of some of the other derivatives. The mass spectrum of the m-dimethylaminobenzylidene (m-DMB) derivative of the ethyl ester of the peptide Val-Ile-Ala (XXX) may serve to illustrate the point. The peptide Val-Ile-Ala has been utilized as a test peptide for numerous derivatives (106, 356, 392) and is therefore rather useful for a comparison:

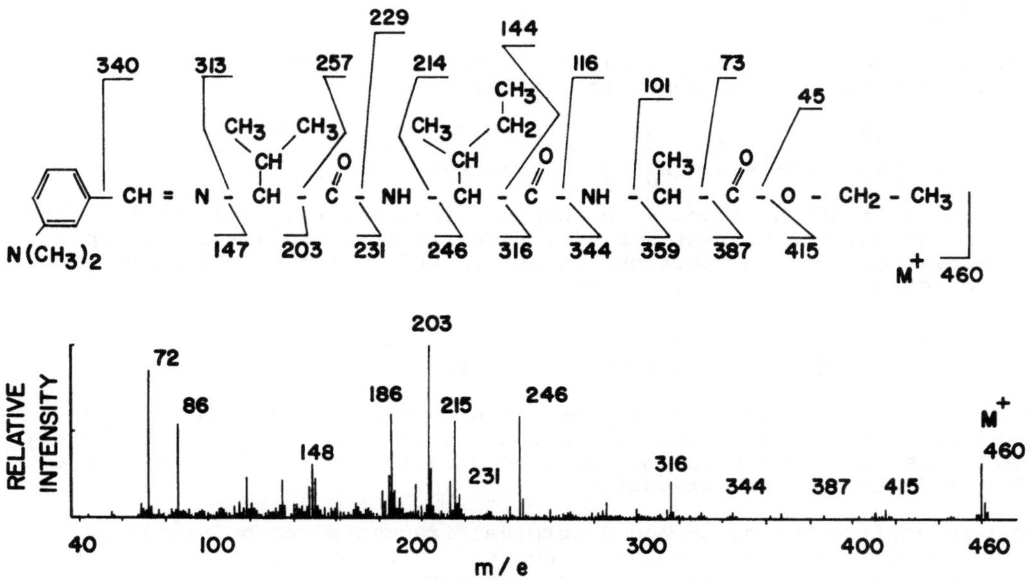

Fig. 8. Mass spectrum of m-DMB-Val-Ile-Ala-OCH_2CH_3 [adapted from Ref. (176)]

It can be seen from Figure 8 that Schiff base derivatives, like their N-acyl counterparts, give rise to peaks due to cleavage at positions A and B (Fig. 5). Peaks due to cleavage in positions A (m/e 203, 316 and 387) together with the intense M^+ peak serve to identify the sequence. Peaks resulting from cleavage in positions B (m/e 231, 344 and 415), while less intense, may serve to confirm the sequence assignment. A significant number of peaks remain in this spectrum. Peaks due to all C-terminal fragments are present in this spectrum but are usually of low intensity and therefore not significant. The only exception to this generality is the relatively intense peak at m/e 215 which is attributed to a C-terminal fragment. Cleavage of the N-C bond of the isoleucine residue and retention of the charge on the C-terminal fragment would produce an ion of m/e 214. It is postulated that this N-C bond cleavage is accompanied by the transfer of a hydrogen from the remainder of the peptide, which then leads to an ion of m/e 215.

However, there remain numerous peaks in this spectrum, some of which are quite intense. Many of these can be attributed to "internal" fragments which contain neither the C- nor the N-terminal blocking group, but may nevertheless contain sequence information.

The intense peaks at m/e 72 and 86 have been attributed to ions (XXXI) and (XXXII), respectively, and while they contain no sequence informa-

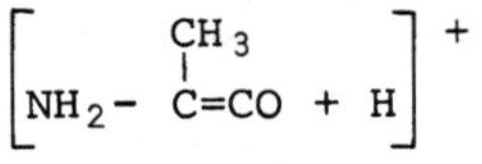

(XXXI)

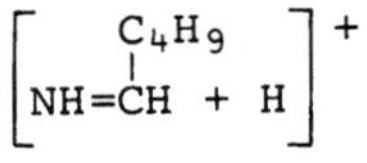

(XXXII)

tion, could be useful in identifying the constituent amino acid. Sequence information is contained in the intense peak at m/e 186 which is attributed to ion (XXXIII).

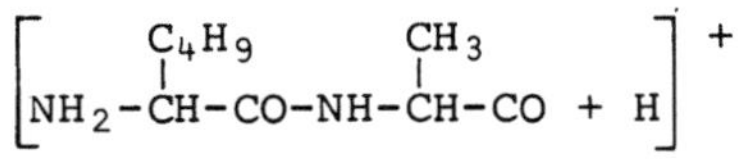

(XXXIII)

Another intense peak at m/e 213 is attributed to ion (XXXIV) which once again contains sequence information.

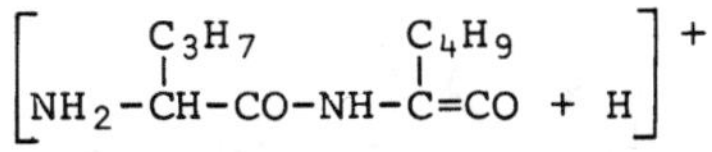

(XXXIV)

The identification of one more internal fragment will suffice for this illustration. The peak at m/e 241, which is still relatively intense for this part of the spectrum, is attributed to ion (XXXV) which once again contains sequence information.

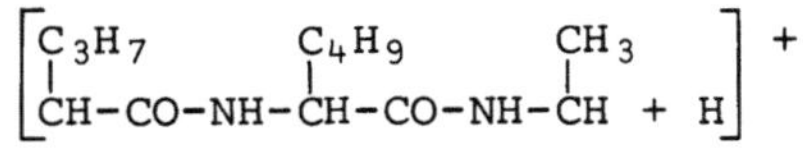

(XXXV)

The internal fragments, once their origin is recognized, aid in the assignment of the sequence. The mechanism by which these internal fragments are generated have not yet been studied and would constitute a fruitful area for investigation.

The identification of C- and N-terminal fragments can be simplified by the use of appropriate isotopic labels. The utilization of a one to one mixture of CD_3OH and CH_3OH for the esterification of the peptide and of, say, p-bromobenzaldehyde for Schiff base formation yields a suitably labeled derivative. Fragments containing the C-terminal ester, but not the N-terminal group, would occur as a doublet of equal intensity three atomic mass units apart. Fragments containing the N-terminal group, but not the C-terminal, would exist as a doublet of nearly equal intensity due to the isotopes of bromine, which have masses of 79 and 81 atomic mass units, respectively. A peptide containing one amino and one carboxyl group would display a quadruplet in the mass spectrum. The increments in mass from the lowest mass peak (Br_{79} and CH_3OH) in such a cluster are two, three and five atomic mass units. The peaks are of nearly equal intensities. The relative intensities and the clusters expected from more complicated combinations, i.e. two carboxyl groups plus one amino group etc., can be readily calculated (43, 45, 72, 175, 297). Any deviation from the unexpected ratios indicates the super-position of another ion and is easily recognized.

Schiff base derivatives of peptide esters are relatively volatile and amenable to gas chromatographic separation. This property is of considerable importance because the direct combination of gas chromatography and mass spectrometry is a powerful tool for structure determination. Day et al. (107) reported that peptides containing arginine, were amenable gas chromatographic separation following treatment with 2-pyridinecarboxaldehyde. It was not necessary to pretreat these peptides by any of the methods discussed previously. The amino acid sequence could be assigned for various peptides regardless of the position of arginine in the chain.

C. Analysis of Peptide Mixtures

The previous discussion demonstrated that the amino acid sequence of individual peptides can be determined from the mass spectrum. Several formal schemes have been proposed for the deduction of the amino acid sequence from the mass spectrum which may proceed either from the C-terminal end of the peptide following identification of the M^+ peak (31, 466) or from the N-terminal relying on a knowledge of the N-protecting groups (46, 400). In the simplest case the interpretation may be carried out by inspection of a low resolution spectrum, while in the most sophisticated schemes the interpretation is accomplished by a computer and relies on high resolution data for all ions in the spectrum (46, 400). The advantages of high resolution mass spectrometry, which allow the identification of all ions in terms of their atomic composition, are in part mitigated by the lower sensitivity of instruments operating in the high resolution mode and the high cost of these facilities which limit their general availability (306, 449).

1. Fractional Vaporization of Peptide Mixtures

When peptides are too large, even after derivatization, to yield good mass spectra, other approaches are needed. In the following discussion we will concentrate on a few selected cases from the literature which demonstrate successful strategies.

Ling et al. (263), in their attempt to confirm by mass spectrometry the amino acid sequence of the luteinizing hormone releasing factor, could not obtain a suitable mass spectrum from a derivative of this substance. The amino acid sequence of the peptide was known to be pGlu-His-Trp-Ser-Tyr-Gly-Leu-Arg-Pro-Gly-NH_2, where the N-terminal residue pGlu is pyroglutamic acid. The following discussion assumes a knowledge of the amino acid composition, but no information concerning the exact sequence except for the N-terminal pGlu residue.

Digestion with the proteolytic enzyme chymotrypsin would be expected to yield a maximum of three fragments from this peptide; two if one of the aromatic residues occurred in the C-terminal position. It would of course be possible to carry out a digestion and to separate the resulting fragment by standard methods such as column chromatography, preparative paper electrophoresis or paper chromatography, but these time consuming steps could be avoided by analysis of the mixture.

Ling et al. first treated the peptide with acetylacetone, by the method of Bacon et al. (28), to convert the guanido group of arginine to a 2,4-dimethylpyrimidyl derivative suitable for mass spectrometry. The derivative was then digested with chymotrypsin and, without separation of the products, acylated with an equal mixture of $(CH_3CO)_2O$ and $(CD_3CO)_2O$. The acylated product was then treated with methanol saturated with HCl and evaporated to dryness. In the final steps the peptide derivatives were permethylated, extracted from the reaction mixture and transferred to the solid insertion probe of a mass spectrometer. Spectra were recorded at low resolution. The probe temperature was gradually raised from 80°C to 300°C and a spectrum was recorded at intervals of approximately 4°C.

The cumulative record of the spectra indicated that a peak at m/e 192 reached a maximum intensity at a vaporization temperature of 128°C. This particular peak could be attributed to a peptide with a C-terminal o-methyltyrosine residue. From doublet peaks at m/e 158 and 161 as well as m/e 130 and 133 the sequence of this peptide was deduced to be Ser-Tyr.

At a probe temperature of 208°C two peaks, one at m/e 98 and another at 215 reached their maximum intensity. The peak at m/e 98 identifies a fragment expected from the N-terminal pyroglutamic acid residue, while the peak at m/e 215 is indicative of a tryptophan residue in the C-terminal position of a peptide. The identification of a doublet at m/e values of 114 and 117 indicated the presence of a N-terminal glycine residue. It was clear, therefore, that the spectrum contained at least two peptides, one with an N-terminal pyroglutamic acid, the other with a N-terminal glycine residue.

The amino acid sequence of the peptide with the N-terminal glycine residue could be determined by searching for additional doublet peaks and comparing these to the mass increments expected from a knowledge of the amino acid composition. No difficulty was encountered in deducing the sequence of this peptide as Gly-Leu-Arg-Pro-Gly-NH_2. The C-terminal amide identifies this fragment as the C-terminal of the original peptide.

The sequence of the peptide with the N-terminal pyroglutamic acid residue could be identified quite readily as pGlu-His-Trp from appropriate nondoublet peaks. Doublet peaks were not expected because the N-terminal pyroglutamic acid residue did not possess a free amino group which could be acylated. No ambiguity was encountered in reassembling the sequence of the original peptide.

The larger the original peptide and the more similar the degradation products, the more likely the prospect of ambiguities in the interpretation of mass spectra of mixtures. Systematic studies by Wipf et al. (469) and Morris et al. (305, 306) led these two groups of researchers to somewhat different conclusions with respect to the analysis of peptide mixtures. Wipf et al. (469), using high resolution mass spectrometry, chemical ionization, metastable ion and fractional vaporization data concluded that many ambiguities would remain when low resolution mass spectrometry is used for the analysis of peptide mixtures. Morris et al. (306), on the basis of their work on peptides derived from the enzyme ribitol dehydrogenase, concluded that low resolution mass spectrometry was only adequate but in fact preferable because of the lower sensitivity, which results from the operation of a mass spectrometer in the high resolution mode.

For practicing biochemists with facilities for chemical fractionation of complex peptide mixtures (column chromatography, electrophoresis, paper chromatography) and access to low resolution mass spectrometry facilities, the approach by Morris et al. (306) should be of particular interest. A combination of classical separation techniques and the analysis of mixtures of peptide derivatives after partial purification might be optimal for laboratories without facilities for the direct coupling of gas chromatography and mass spectrometry. The utilization of dual-purpose reagents which combine the detection of peptides on

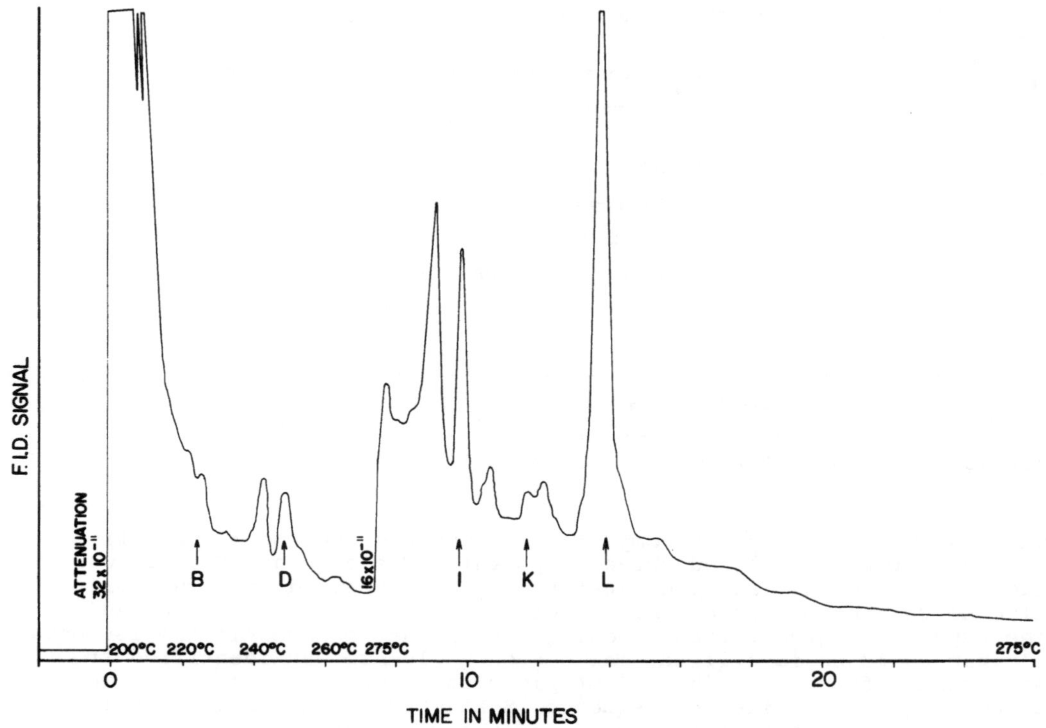

Fig. 9. Gas chromatogram of a mixture obtained by partial acid hydrolysis of Cbz-Ile-Glu-Asp(SBzl)Cys-Pro-Leu-GlyNH$_2$ and subsequent derivatization (see text). [Adapted from Ref. (214)]

chromatograms and their derivatization for mass spectrometry (146) should be advantageous in such an approach.

2. *Direct Coupling of GC-MS for Analysis of Peptide Mixtures*

For laboratories with facilities for the direct coupling of gas chromatography and mass spectrometry other approaches might be considered. Jayasimhulu and Day (214) could not obtain a suitable mass spectrum from the peptide derivative Cbz-Ile-Glu-Asp-(S-Bzl)Cys-Pro-Leu-GlyNH$_2$ by direct introduction with the solid insertion probe. One μmol of this peptide was subjected to partial acid hydrolysis by suspension in 0.5 ml of 6N HCl and heating the mixture for 20 minutes at 105°C in a sealed, evacuated tube. The HCl was then removed by flash evaporation and the partial hydrolysate esterified with a 1:1 mixture of CH_3OH and CD_3OH and following removal of the excess alcohol treated with p-bromobenzaldehyde in glacial acetic acid. About half of the final mixture, in the form of an acetic acid solution, was used for analysis by GC-MS. The gas chromatogram is shown in Figure 9; the letters indicate the positions where mass spectra were taken. Mass spectra taken in these positions were sufficient to reconstruct the amino acid sequence, other mass spectra would serve to confirm these assignments. Table 2 shows the partial sequences which could be deduced from the mass spectra and their assembly by overlapping fragments to the sequence of the entire peptide. It is clear that the gas chromatographic separation of the

Table 2. Reconstitution of the amino acid sequence of a heptapeptide

GC peak (Fig. 9)	Amino acid sequence deduced from mass spectrum
L	ILE-GLU
L	GLU-ASP
D	ASP-(S-Bzl)CYS-PRO
L	(S-Bzl)CYS-PRO
K	PRO-LEU
I	LEU-$GLYNH_2$
B	$GLYNH_2$
Total sequence	ILE-GLU-ASP-(S-Bzl)CYS-PRO-LEU-$GLYNH_2$

Adapted from K. Jayasimhulu and R.A. Day (214).

various fractions was not complete. For instance the peak identified by the letter L contained three components. The sequence Ile-Glu was identified from an isotopic cluster of peaks expected from the derivative containing one amino and two carboxyl groups. The M^+ peaks and the expected sequence peaks showing appropriate changes in the isotopic cluster could be identified. The sequence Glu-Asp was identified from an internal fragment with an isotopic cluster characteristic of two carboxyl groups, while the sequence (S-Bzl)Cys-Pro could be identified from an internal fragment showing no isotopic cluster but the characteristic loss of 91 mass units due to the elimination of the benzyl protecting group. The simplicity and speed of the chemical derivatization and relative ease with which the sequence could be deduced from relatively few spectra are appealing features of this approach.

Most degradation procedures, such as partial acid hydrolysis and hydrolysis by mixtures of various endopeptidases, generate peptides of various sizes. In contrast, dipeptidylaminopeptidases (DAP) degrade proteins from the N-terminus by the successive removal of dipeptides. DAP I hydrolyses peptides from the N-terminus unless the bond to be cleaved involves a proline residue or if the poly-peptide chain contains an N-terminal arginine or lysine residue. DAP IV hydrolyses dipeptides from the N-terminus if the bond to be cleaved involves the carboxyl group of proline but not if it involves the amino group of proline. Caprioli and Seifert (84) demonstrated recently that a mixture of DAP I and DAP IV generates dipeptides from the N-terminus of a protein until proline appears in the third position of the remaining peptide chain. The dipeptides derived from the proteins were esterified and N-acylated and then subjected to analysis by GC-MS. Overlapping fragments of dipeptides can be generated by subjecting an aliquot of the protein to one round of degradation by the Edman procedure, followed by degradation with the mixture of DAP I and DAP IV. GC-MS analysis of the dipeptides formed by treatment with these enzymes offers an alternative approach to sequence analysis.

The most highly developed methodology for the analysis of peptide mixtures involves the polyaminoalcohol derivatives (319-323). The recent determination of the primary structure of the carboxypeptidase inhibitor from potatoes, a polypeptide of 39 amino acid residues, serves to illustrate the potential of this approach (180, 321).

A 1.0 µmol sample of the S-carboxymethylated inhibitor was subjected to partial acid hydrolysis with 2 ml of 6N HCl at 110°C for 35 min in an evacuated and sealed tube. The acid was removed and the peptide mixture esterified trifluoracetylated, reduced with $LiAlD_4$ and o-trimethylsilylated as shown earlier (XXIV → XXV). The derivatives were dissolved in 50 µl of benzene and 8% of the sample injected into the GC together with a 0.5 µl solution of alkane standards. Three sets of data were generated (a) mass spectra were recorded every 4.7 s, (b) selected ion records were obtained of all ions between m/e 30 and 755, (c) retention indices were assigned to all mass spectra. From the mass spectral data, combined with the retention indices, the sequence of the various peptide fragments were determined.

A total of 52 peptides were identified in the partial acid hydrolysate together with some free amino acids particularly serine, threonine, aspartic acid, glutamic acid and carboxymethylcysteine. The occurrence of free amino acids is indicative of the fact that the hydrolysis time might already have been too long. The authors suggest that the hydrolysis time should be reduced to 20 min or less. Several peptide bonds were not represented in the 52 oligopeptide, particularly those involving the acid labile bond of serine and threonine.

A second hydrolysis experiment was therefore conducted using a mixture of endopeptidases consisting of thermolysin, chymotrypsin, elastase and trypsin. This mixture of proteolytic enzyme could be expected to produce numerous peptide fragments but few amino acids. The analysis of the mixture following derivatization indicated the correctness of the assumption. Most importantly the acid labile peptide bonds involving serine and threonine were represented in this mixture.

Selected ion records play an important part in the analysis of such complicated mixtures. From the expected fragmentation patterns of polyamino alcohols, the various fragments can be predicted. For instance, a peak at m/e 291 corresponds to the cleavage at A_2 of peptides containing the sequence Ala-Asp or Asp-Ala (see Fig. 7). A peak at m/e 291 emerged at scan 81. The remainder of the spectrum indicated that the sequence was Ala-Asp. Another peak at m/e 291 was observed at scan 172. From the retention index it was clear that this peptide derivative had to be that of a tri- if not a tetrapeptide. Checking the selected ion record for all A_3 ions by the addition of 44 plus the masses of all possible side chains showed that a peak at m/e 376 reached a maximum also at scan 172, which indicated that the third residue was proline. However, the calculated retention index for Ala-Asp-Pro is lower than the one observed, and therefore a search was made for a fourth residue which turned out to be Ile.

Selected ion records and retention indices were particularly important in the identification of unresolved peaks containing two or more peptides. When all of the peptides identified from the acid hydrolysate and the enzymatic digest were combined, two sequences differing in the position of one Asp residue were obtained from the computer output. A search was then made for tripeptides which would conclusively differentiate between the two sequences. The identification of a peptide of the sequence Asp-Asp-Cys confirmed one of these. The peptide had been missed in the original search because the derivatives had undergone a reaction eliminating TMS from the second Asp side chain. This reaction was known, but had usually been observed to give rise to only a minor product.

In summary, it was possible to obtain an unambiguous amino acid sequence of 39 residues from a knowledge of the amino acid composition and from two separate hydrolysis experiments. High resolution mass spectrometry was not required for the analysis. However, facilities were required for the direct coupling of gas chromatography, mass spectrometry and the computer aided collection and evaluation of the data.

IV. Concluding Remarks

A. Recent Developments in Instrumentation

Most of the discussion of mass spectrometry and its application to sequence determination involved the classical techniques of electron impact ionization and the introduction of samples into the mass spectrometer by means of a solid probe or by direct coupling to a gas chromatograph. Considerable attention has been given to the classes of volatile derivatives which have attained relatively wide acceptance or which were deemed particularly promising. This emphasis is justified because the overwhelming volume of work reported involves these derivatives and the basic instrumentation required for their analysis is generally available. Nevertheless, advances in instrumentation might eventually reduce or eliminate the need for the chemical derivatization of peptides and increase the molecular weight range which can be analyzed by mass spectrometry. Winkler and Beckey (468) using the technique of field desorption demonstrated that peptide derivatives generally regarded as highly involatile can give molecular ion peaks. A case in point is the tetrapeptide derivative Ac-Gly-Arg-Arg-GlyOCH_3 for which a very intense M^{+}+1 peak could be obtained. Beuhler et al. (41) using a rapid heating technique, i.e., a sample evaporation time of a few seconds, and gentle ionization of the neutral gaseous peptides with NH_4^+ obtained spectra of underivatized peptides, which would have been quite unattainable by conventional means. MacFarlane and Torgerson (272) using a ^{252}Cf plasma desorption technique to volatilize and ionize samples obtained M^{+} peaks from underivatized gramicidins at m/e values of 1904 and 1918, respectively.

Improvements in the sensitivity of mass spectrometers might allow the analysis of much smaller samples or alternatively allow the detection of ions of very low abundance. McIver et al. (195) have developed a trapped ion cyclotron mass spectrometer that can detect vapor pressure in the range of 10^{-10} Torr, a factor of 10,000 better than conventional electron impact mass spectrometers with a solid inlet probe.

Improvements in the methodology for combining liquid chromatography and mass spectrometry offer promise for the application of chemical ionization mass spectrometry to sequence determination (23, 29).

B. Relationship of Mass Spectrometry to Conventional Methods for Sequence Analysis

Mass spectrometry in the opinion of several authors (206, 317) should not be considered as a tool to replace the existing methods for sequence analysis, especially the automated Edman degradation, but as a means of obtaining complementary sequence information. This view holds that each method taken by itself is subject to certain limitations which are not the same because of the fundamental differences in methodology. The optimal approach to sequence analysis of large proteins would then be seen as one where both approaches are used simultaneously. The combination of results of two independent methods could be considered much more reliable than those generated by either method alone.

Peptide Sequence Analysis by Nuclear Magnetic Resonance Spectroscopy

E. NIEBOER and H. FALTER

I. Introduction

Nuclear magnetic resonance spectroscopy (NMR) has found wide application to biochemical problems. This is reflected in the number of books (61, 130, 212, 243, 473) and review articles (95, 96, 163, 169, 294, 300, 325, 375) that describe this area of research. In recent years, considerable progress has been made in studies of the structure in solution of peptides, proteins, nucleotides and other biomolecules. Major technical developments were responsible for this success. These include: (1) the manufacture of superconducting magnets, (2) the general availability of Fourier transform NMR spectrometers, and (3) the discovery of paramagnetic chemical shift and relaxation enhancement (broadening) agents. In tandem, these advancements have extended the sensitivity and scope of NMR spectroscopy considerably. As a consequence, carbon-13 resonance studies no longer require an isotopic enrichment step (257, 420), and the available resolution has been improved to such a level that even the details of multiplet structure in protein ^{1}H-spectra may be discerned (77-79, 130). The paramagnetic chemical shift and broadening probes may be used to simplify complex NMR spectra, and are useful in making peak assignments. Roberts and Jardetzky (375), in 1970, concluded that sequence determinations by NMR would likely be of limited value because of the inability to resolve adquately the resonances of ^{1}H nuclei of the various amino acid residues. In this chapter, we will show that advances in NMR technology have generated renewed interest in peptide sequence analysis, and that this application of NMR has considerable merit and promise.

II. NMR Principles

The basis principles of NMR are well documented in standard texts (60, 130, 141, 211, 212, 473). We will review only those aspects essential to the understanding of the contents of this chapter.

The position of resonances in a spectrum are measured relative to some internal (or external) standard (Fig. 1). ^{1}H chemical shifts are usually referred to tetramethylsilane (TMS) for measurements in organic solvents, and to 2,2-dimethyl-2-silapentane-5-sulfonate (DSS or TSS) in aqueous media. These same standards are employed in ^{13}C work, although the measured chemical shifts are often expressed relative to CS_2. Chemical shifts, δ, are best expressed in parts per million (ppm), a dimensionless unit independent of the applied magnetic field: $\delta = 10^6 (\nu_{ref} - \nu_i)/\nu_o$. Here ν_i is the resonance frequency of the observed nucleus, ν_{ref} that of the reference compound, and ν_o the operating frequency of the spectrometer. The range of ^{13}C shifts is considerably larger ($\sim$200 ppm) than those for hydrogen nuclei ($\sim$12 ppm).

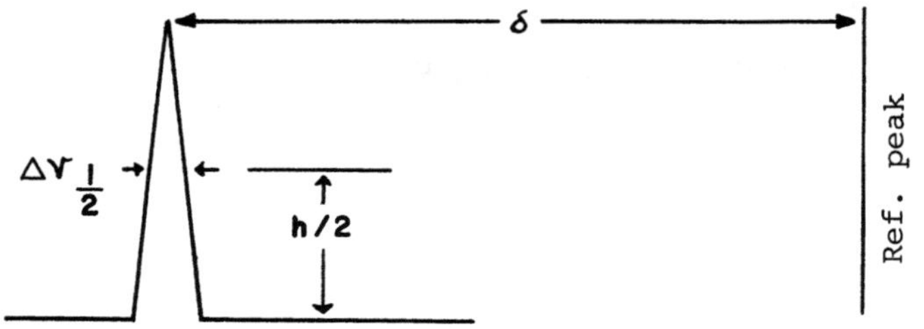

Fig. 1. Definition of the chemical shift, δ, and line width, $\Delta\nu_{1/2}$ in Hz, at half height (h/2)

The line width at half height of the resonance signal, $\Delta\nu_{1/2}$, is defined in Figure 1. It is related to the spin-spin (transverse) relaxation time, T_2, by the equation: $1/T_2 = \pi\Delta\nu_{1/2}$. T_2 is the lifetime or phase memory time of a nuclear spin state. A related parameter T_1, the spin-lattice of longitudinal relaxation time, is also useful in modern NMR measurements. This time constant characterizes the return to equilibrium with its surrounding (the "lattice") of a perturbed nonequilibrium spin system such as that subject to a pulse of the resonance radio frequency. It is measured by pulse NMR methods (130, 149, 212, 473).

Both ^{1}H and ^{13}C have nuclear spin state quantum numbers of ½. The hydrogen nucleus has a 99.98% natural abundance. This is in contrast to the 1.1% abundance of ^{13}C. Low abundance combines with low natural sensitivity to NMR detection in making ^{13}C a difficult nucleus to detect. However, Fourier transform NMR (vide infra) has met this challenge. ^{1}H spectra have multiplet structure due to the proximity of other ^{1}H nuclei. The separation between two lines split by such coupling is the spin-spin coupling constant J. Because of the low natural abundance of ^{13}C in compounds, ^{13}C-^{13}C spin-spin coupling is normally not detected. Routinely, ^{1}H-^{13}C coupling is "noise decoupled" when recording a ^{13}C spectrum. Consequently, ^{13}C spectra are relatively uncomplicated.

The continuous wave mode (CW) has been the most common manner of operating NMR spectrometers. It normally consists of slowly sweeping the applied magnetic field while subjecting the sample to a fixed radio frequency (rf). Fourier transform NMR consists of applying an intense pulse of rf energy to the sample for a very short time, and the measurement of the resulting free-induction decay signal from the nuclear spins in the sample (149, 212, 473). The sample is irradiated with a band of rf frequencies so as to excite all the nuclei in the sample simultaneously. The free-induction decay signals of repeated pulses are readily accumulated, and the sum is Fourier-transformed by a computer to give the conventional CW spectrum. This pulsed Fourier transform technique makes possible the study of less sensitive nuclides such as ^{13}C as well as allowing the use of dilute solutions in ^{1}H measurements. In addition, the free-induction decay signal is readily manipulated mathematically to give improved resolution. This procedure further permits the convenience of substracting spectra (77-79).

Finally, deuterated solvents are essential in ^{1}H NMR. The need for this is obvious. For example, hydrogen nuclei have a concentration of 110 M in water, and the corresponding signal would swamp out the signals of the compound under study. Several instrumental methods are also available for repressing solvent peaks in FT measurements (50, 80, 212, 473).

III. Amino Acid Composition and Residue Identification

A. Analysis of High Resolution ^{1}H Spectra

The proton[1] resonances of the common free amino acids are observed in the spectral region DSS to -8 ppm (downfield). Tabulations of the chemical shifts and J values are available (61, 130, 212, 293, 375, 401, 473) for the various ionic forms of the amino acids. Characteristically, the resonances corresponding to the aromatic side-chain protons of His, Phe, Trp, and Tyr are well separated from backbone and other aliphatic side-chain signals and occur in the range -7 to -8 ppm relative to DSS, while the corresponding aliphatic resonance signals are found 1 to 4.5 ppm downfield from this reference. The incorporation of amino acids into "random coil" peptides affects the chemical shift in a predictable fashion. For example, the formation of a peptide bond involving the amino group of glycine produces a downfield shift of 0.66 ppm in the αCH protons (312, 375). Peptide bond formation with the carboxyl group is accompanied by a smaller shift. Incorporation of zwitterionic forms into peptides is accompanied by similar spectral changes (275). With exception of the titration of the terminal amino or carboxyl functional groups, there appear to be no significant primary structure effects on the resonance positions of side-chain protons. In addition, examination of oligoglycines (310, 375) and glycine-containing dipeptides (312, 375) has indicated that the nature of the nearest neighbor has little effect on αCH resonances, unless helix formation is induced under specific solvent and temperature conditions (310). Consequently, in the absence of secondary and tertiary structural effects, the spectrum of a polypeptide is a superposition of the individual spectra of its constituent amino acids, provided corrections are made for the predictable and additive shift perturbations due to titration and peptide bond formation (312, 375). Considerable success has been achieved in simulating spectra of random coil peptides and denatured proteins on the basis of such "corrected" chemical shifts of the appropriate amino acid residues (73, 212, 293, 473). It should be noted that studies of spin-spin coupling constants do indicate the occurrence of some secondary structure for short random coil peptides containing residues with bulky side chains (73, 473). Certain side chains (e.g., aromatic rings of Phe, Tyr and Trp) would appear to assume preferred spatial orientations.

It is important that we examine the titration effects in greater detail. Sheinblatt (401) was successful in sequencing di- and tripeptides by examining the spectra of the cationic (+), zwitterionic (±), and anionic (-) forms. The αCH ionization shifts for amino acids were found to be ∿0.4 ppm downfield for $-COO^{-} \rightarrow -COOD$, and ∿0.5 ppm upfield for $-\overset{+}{N}D_3 \rightarrow -ND_2$, while they were ∿0.3 and 0.5-0.6 ppm, respectively, for the αCH protons of the ionizing residues in peptides (37, 375). Nakamura and Jardetzky (310), and, more recently, with better resolution Anteunis and Callens (19), have demonstrated that the perturbations due to protonation of the terminal amino function may be discerned even for the second-residue α-protons of pentaglycine. Similarly, the C-terminal protonation also induced a detectable chemical shift in the α-protons of the fourth residue. Ionization shifts can also be detected for side-chain protons, especially those of the terminal ionizing residue (37, 375, 401, 473) (the chemical shift changes due to protonation of side-chain functional groups are considered below.) The sequence position of four residues of a peptide may thus be identified

---

[1] This term is widely used in NMR to represent the hydrogen nucleus.

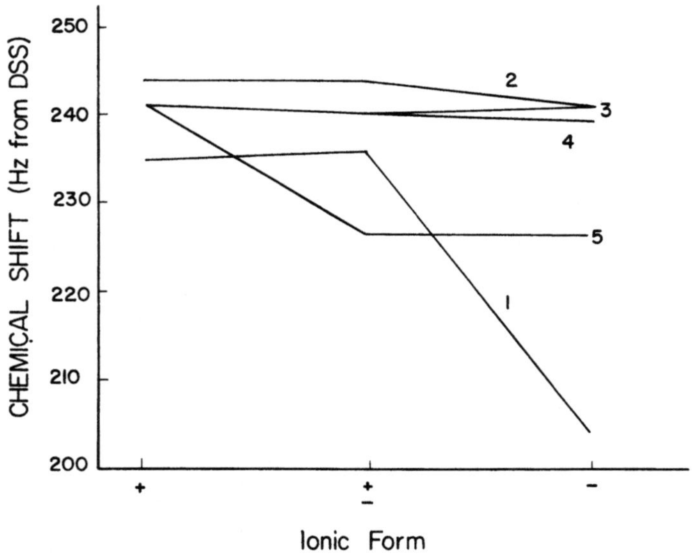

Fig. 2. Chemical shifts of the methylene protons for the various forms of pentaglycine. The residues are numbered 1 to 5 from the N- to the C-terminus. Data taken from Nakamura and Jardetzky (310)

by examining the pH dependence of its proton shifts. This approach to sequencing has been designated as the Sheinblatt method (19), and is illustrated in Figure 2.

The identification of a residue of known sequence position may be achieved by employing standard NMR techniques such as measurements of chemical shifts, relative peak areas, and coupling constants. Double-irradiation decoupling techniques (130, 212, 473) are very useful in making definite assignments. Proton spectra of amino acids and peptides may be divided into four general spectral regions downfield from DSS: methyl protons, ~1 ppm; $\beta(\gamma,\delta,\varepsilon)$CH, 1.5-3 ppm; αCH, 3.5-4.5 ppm; and aromatic protons, 7-8 ppm (2, 9, 26). This classification was established for D_2O but would also appear to be valid for other polar solvents such as d_6-DMSO (75, 473) and methanol (vide infra). Identification of peaks is further aided by shift changes accompanying the ionization of side-chain functional groups. This is considered in the next few paragraphs.

The aromatic region has contributing resonances from His, Phe, Trp, and Tyr (92, 94, 95, 375). Both the C-2 and C-4 protons of the imidazole unit have chemical shifts that are very pH-sensitive. This is illustrated in Figure 3 for the dipeptide His-Phe (94) and for the free amino acid in Figure 4 (387). The pK_a value of the imidazole nitrogen in histidine may be evaluated directly from titration curves like those in Figure 4, and was found to be 6.1 at 22°C compared to the range of 6-7 log units observed in peptides (94). (Interestingly, the pK_a values observed for this functional group in native proteins vary widely from 5-8 log units (280), and thus the imidazole is a good reporter group of secondary and tertiary structure.) The ionization of the functional group of a tyrosine residue may also be followed by the chemical shift changes of the corresponding aromatic ring protons at pH values above 8.5 (93, 375).

Among the nonaromatic side-chain residues, the following have ionizing functional groups: Asp, Glu, Lys, Arg and Cys. Some care is needed to distinguish side-chain carboxylic acid and amino functions from the terminal ones. However, the pK_a values of the side-chain groups differ by at least one log unit from those of the α-position in both amino

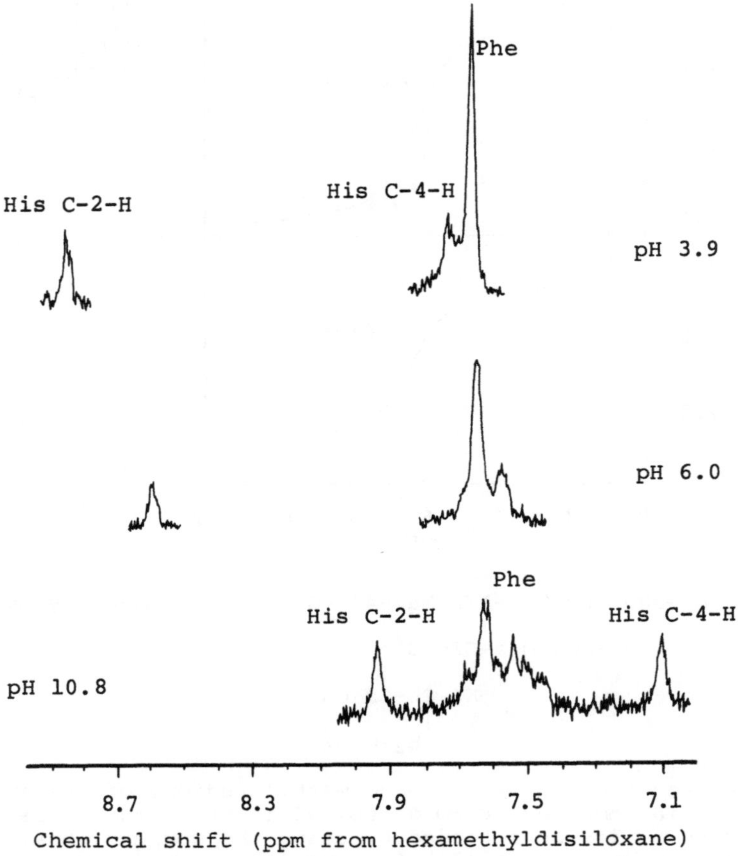

Fig. 3. Proton spectra of the aromatic region of the dipeptide His-Phe at three different pH values. Adapted with permission from Cohen (94)

acids (274) and peptides (63, 401). Consequently, pH-dependent shifts due to side-chain carboxylate and amino group deprotonation occur at higher pH values than those associated with α-group ionization. The mercapto group of cysteine is also amenable to titration, and both pH- and oxidation-induced shifts should help in the identification of this residue (375). The side chain of Arg is expected to be protonated under most experimental conditions (pK_a ∿12.5 for amino acid).

The revision of the primary sequence of alamethicin illustrates the potential of NMR spectroscopy in sequence analysis. A detailed NMR analysis (281) has indicated that this cation transport peptide is linear rather than cyclic, and, in addition to the 18 amino acids determined previously by conventional chemical techniques (41) and mass spectrometry (333), contains two new end-groups. One of these, β-phenylalaninol, is attached to the γ-carboxyl group of the only glutamic acid residue in the peptide, while the other, N-acetyl-2-methylalanine (Ac-Mea), is the first amino acid residue in the sequence.

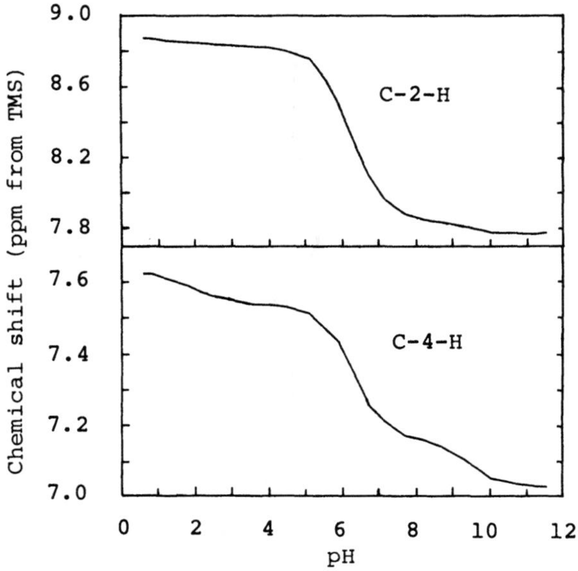

Fig. 4. Chemical shifts of the C-2 and C-4 imidazole protons of L-histidine as a function of pH. Adapted with permission from Sachs et al. (384)

```
Ac-Mea-Pro-Mea-Ala-Mea-Ala-Gln-Mea-Val-Mea-Gly-Leu-Mea-Pro-

     Val-Mea-Mea-Glu-Gln
                     |
                     NH-CH-CH2OH
                        |
                        CH2-C6H5
```

The aromatic, $\alpha CH + \beta CH_2$, and methyl regions of the spectrum of alamethicin component F 30 dissolved in dry methanol is reproduced in Figure 5. It was recorded by using the computational convolution enhancement technique (79) in conjunction with Fourier transform mode operation of a 270 MHz spectrometer. The approach and evidence used by Martin and Williams (281) in determining this new sequence is now summarized:

1. anomalous peaks in the spectrum (marked Mea-O, phenylalaninol (Pal), and aromatic) were inconsistent with the previously published sequence

Fig. 5a-c. Proton spectrum of alamethicin component F 30. The solvent was methanol (nominal pH 1.5), and the spectrum was obtained with convolution enhancement (79). Some of the most important regions of the spectrum are labeled. (a) 7.4-8.7 ppm covers the NH region, which includes resonances of the phenylalaninol (Pal) and the extra amino acid that was found and is labeled Mea-O to show that it occurs before the first amino acid, proline-1, of the chemically determined sequence. Between 7.0 and 7.3 ppm the aromatic resonances are seen. (b) Various CH and CH_2 groups are observed, and again several of them are labeled as belonging to specific residues. These assignments are based on spin-decoupling experiments (the coupling is clearly seen in the spectrum), and on probe experiments with lanthanides and chromicyanide (see Sect. III.C below). The "acetyl" peak and the phenylalaninol and Mea-O peaks (including the aromatic peaks) are all quite inconsistent with the previously reported structure. (c) Tetramethylsilane is a marker. Numbering of the residues is that of the previously determined sequence (see text). Adapted with permission from Martin and Williams (281)

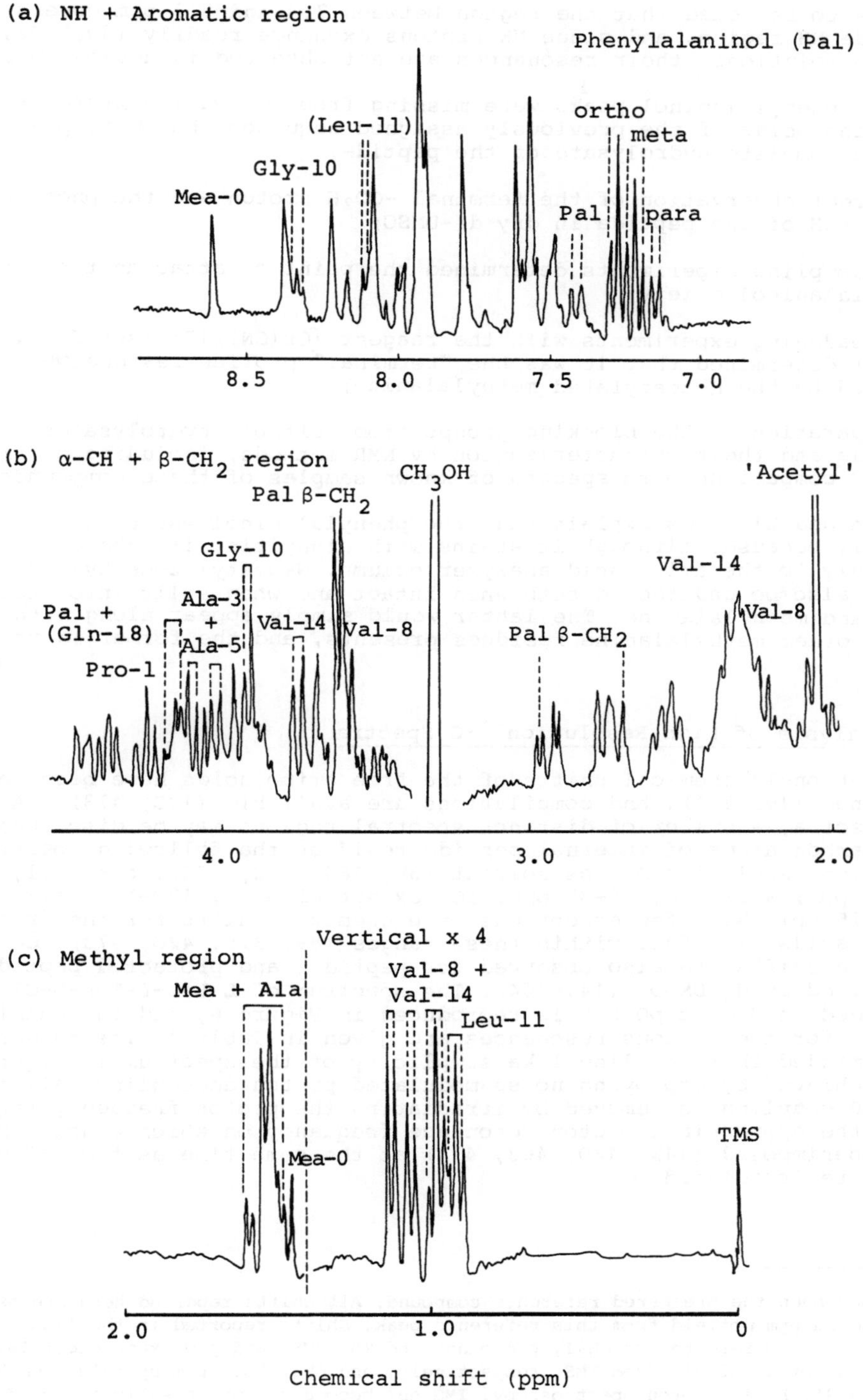
(a) NH + Aromatic region
Phenylalaninol (Pal)
ortho
meta
(Leu-11)
Gly-10
Mea-0
Pal
para
8.5
8.0
7.5
7.0
(b) α-CH + β-CH2 region
CH3OH
'Acetyl'
Pal β-CH2
Gly-10
Val-14
Ala-3
+
Ala-5
Pal +
(Gln-18)
Pro-1
Val-14
Val-8
Val-8
Pal β-CH2
4.0
3.0
2.0
Vertical x 4
(c) Methyl region
Val-8 +
Val-14
Mea + Ala
Leu-11
Mea-0
TMS
2.0
1.0
0
Chemical shift (ppm)

Fig. 5 a-c

(it is to be noted that the region between 7.4 and 8.7 ppm covers the peptide NH region, and since NH protons exchange readily (80, 402, 403) in D_2O solutions, their resonances are not observed in aqueous medium);

2. the phenylalaninol peaks were missing from an equal mixture of the 18 amino acids of the previously assigned sequence, but were present in the complete hydrolysate of the peptide;

3. direct observation of the terminal $-CO_2H$ proton and the phenylalaninol -OH of the peptide in dry d_6-DMSO;

4. decoupling experiments determined the point of attachment of the phenylalaninol moiety;

5. broadening experiments with the reagent $[Cr(CN)_6]^{-3}$ (see Sect. III.C) determined that it was the "terminal" proline residue which was blocked by the N-acetylated methylalanine;

6. separation of the blocking groups from suitable hydrolysates of the peptide and their characterization by NMR methods, including a "fingerprint" comparison with spectra of known samples of these compounds.

Martin and Williams explain that the phenylalaninol was undetected previously because, although it stains with ninhydrin, it adheres very strongly to the amino acid analyzer column. N-acetyl-2-methylalanine would also be undetected both when intact and when split into acetic acid and methylalanine. The latter would simply appear along with the seven other methylalanine residues presents, and the former is volatile.

B. Analysis of High Resolution ^{13}C Spectra

The carbon-13 chemical shifts of the free amino acids have been determined (199-201), and compilations are available (130, 473).[2] As for ^1H spectra, a number of distinct spectral regions may be discerned. For carbon atoms of internal peptide residues the following spectral divisions apply for D_2O as solvent (89, 164, 172, 473): carbonyl, 17-22 ppm; aromatic, 55-85 ppm; αC (except of Gly), 130-145 ppm; βC, 160-175 ppm. With few exceptions, the chemical shifts for the free amino acids also fall within these ranges (89, 375, 420, 473). Comparable shifts are also observed for peptides and protected peptides dissolved in d_6-DMSO (114, 164). The spectrum of L-Leu-L-Ser-L-Glu recorded in H_2O at pD 8.5 is reproduced in Figure 6, and the chemical shifts for the various resonances are given in Table 1. The reader is reminded that the line-like simplicity of the spectrum in Figure 6 was achieved by employing noise-modulated proton decoupling. All the ^{13}C-^1H coupling is removed by irradiating the proton frequency region with the appropriate proton resonance frequency on which a noise band is superimposed (143, 420, 463, 473) at the same time as the ^{13}C nuclei are irradiated.

---

[2] CS_2 has been the preferred reference compound. All shifts reported here are expressed in ppm upfield from this reference peak. Shifts reported in the literature also are often referred to CH_3I, p-dioxane and TMS. CS_2 and p-dioxane occur 194 ppm and 67.4 ppm downfield from TMS, respectively, and CH_3I 20.5 ppm upfield from TMS (420, 473). As in proton spectroscopy, TMS has become the universally accepted standard in ^{13}C measurements (420, 473)

Table 1. ^{13}C chemical shifts of L-leucyl-L-seryl-L-glutamic acid (see Fig. 6)[a]

Carbon nucleus	Chemical shift (ppm upfield from CS_2)
Leucine	
α-CO carbon	16.7
αC	140.0
βC	150.2
γC	168.6
δC	170.7 171.2
Serine	
α-CO carbon	22.1
αC	137.6
βC	131.5
Glutamic acid	
α-COOH carbon	14.5
αC	137.6
βC	164.3
γC	158.9
γ-COOH carbon	11.0

[a]From Gurd et al. (172). Peptide concentration was 0.4M at pH 8.50 (measured at 25°C), and the probe temperature was 36° ± 3°.

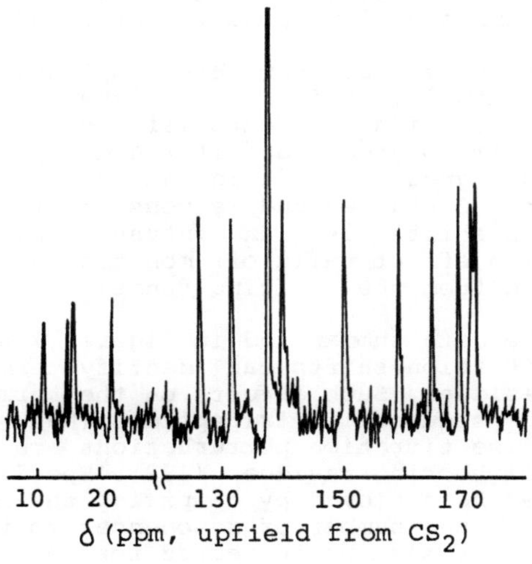

Fig. 6. Natural abundance carbon-13 spectrum of the tripeptide L-Leu-L-Ser-L-Glu at pH 8.5. For details of the experimental conditions and individual peak assignments see Table 1. Redrawn with permission from Gurd et al. (172)

The effect of peptide bond formation on the chemical shifts of free amino acids at pH 6 is summarized in Table 2.

Table 2. Effect of peptide bond formation on chemical shifts[a]

Position of residue in peptide	Shift difference (ppm)[b]			
	carbonyl	αC	βC	γC
N-terminal	4 to 6	1 to 2[c]	-1 to +1	-0.5 to +0.5
Internal	1 to 2	0 to 2	0 to -1[d]	-0.2 to +0.2
C-terminal	-3 to -4	0 to -2	-1 to -2	0 to -0.8

[a]Data were obtained from (172) and (89).

[b]Shift difference in ppm relative to the free amino acid in zwitter-ionic form (pH 5-6). The minus sign denotes a downfield shift.

[c]Excludes Phe bound to Pro when shift is 3 to 4 ppm.

[d]Excludes Pro for which the cis and trans isomers can be distinguished: cis, -0.7; trans (+2.1), the most predominant isomer (164).

The largest shifts occur for the carbonyl carbons of the terminal residues. Small effects are observed for the incorporation of amino acids as internal residues. Grothwohl and Wuetrich (164, 473) have also demonstrated that the nearest-neighbor effects of internal peptide units are small. They examined the protected peptide TFA-Gly-Gly-L-X-L-Ala-OCH_3, where X stands for the 20 common amino acids for studies in d_6-DMSO, and 10 common residues for studies in D_2O. The following nearest-neighbor perturbations were observed for the aqueous study: 0.1-0.7 ppm for carbonyls, and 0.2-0.5 for αC and βC. Similar results were obtained for the less polar solvent d_6-DMSO. Cis and trans (relative abundance 1:4, respectively) isomers could be observed for the peptide with X as proline, as in the corresponding ^1H studies (73).

Protonation titration shifts have also been used to determine the primary sequence of di- and tripeptides by ^{13}C NMR spectroscopy (89). This is illustrated in Figure 7 for the tripeptide L-Leu-L-Ser-L-Glu. A summary of various titration effects is provided in Table 3.

It is evident from this compilation that the ionization shifts for terminal peptide residues are similar to those observed in the free amino acids. One surprising observation is that the deprotonation of an amino group, and it would appear also the mercapto function, induces a greater shift in the βC than the αC. This is a general phenomenon in ^{13}C studies and is consistent with recent theoretical treatments of ^{13}C shifts (34). In contrast, for carboxylate deprotonation the titration effect falls off monotonically with distance of the observed carbon from the ionizing functional group (see Fig. 8).

The data summarized in Figures 7 and 8 and Table 3 leave no doubt that titration shifts can identify terminal residues and ionizing internal residues. Such effects on the terminal residues suffice to identify the sequence in tripeptides (89). Compared to ^1H NMR spectroscopy, these titration perturbations are not as effectively transmitted to neighboring residues (172). Specific assignments and identification can be achieved by comparing the magnitude of chemical shifts and the type and number of resonances to those for amino acids. Of course, the latter must be corrected for the rather small nearest neighbor and somewhat larger titration and peptide formation effects. The ability of observing directly the carbonyl unit is a definite advantage over ^1H NMR spectroscopy. Residue identification is further aided by double

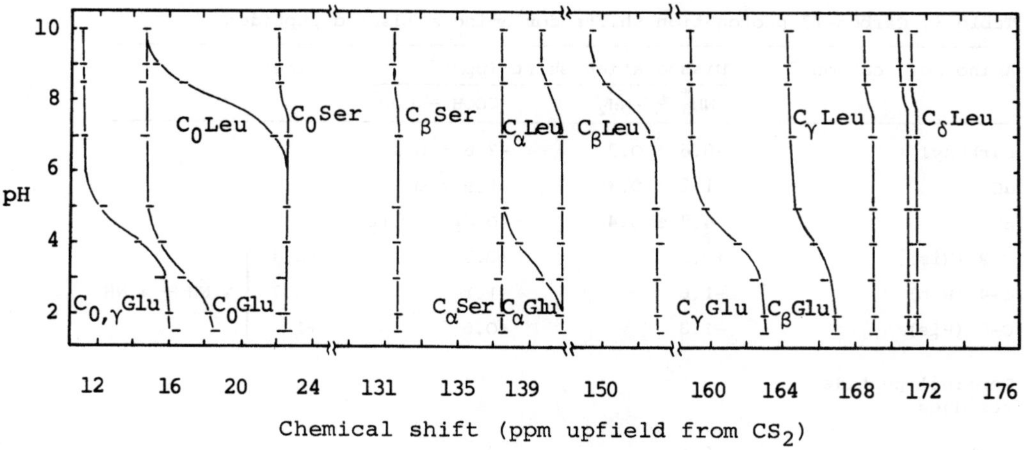

Fig. 7. The pH-dependence of the ^{13}C chemical shifts for the tripeptide L-Leu-L-Ser-L-Glu. C_O, α-carboxyl carbon atom; $C_{O,\gamma}$, γ-carboxyl carbon. Redrawn with permission from Gurd et al. (172)

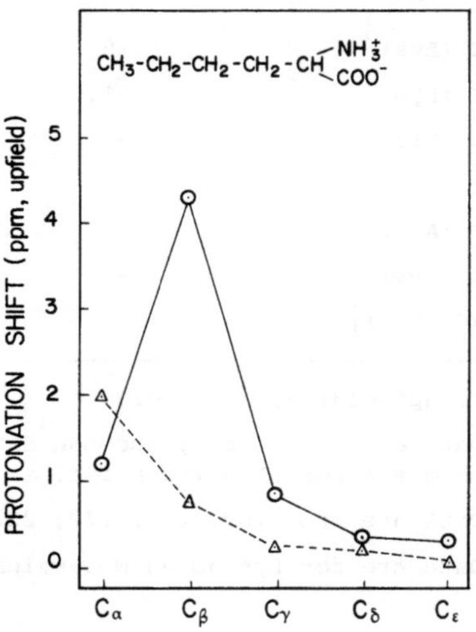

Fig. 8. Upfield shifts of norleucine ^{13}C resonances due to protonation of the amino (⊙) and carboxylate (Δ) groups. Data taken from Batchelor et al. (34)

resonance techniques (130, 212, 473) including selective and off-resonance decoupling to identify specific carbon atoms (10, 143, 164, 360, 463, 464). Considerable success has been achieved in making complete assignments and simulating spectra of short (<10 residues) and intermediate length (10-30 residues) peptides. These include: a pentapeptide comprising the five N-terminal residues of sperm whale myoglobin (172); glutathione (222); amino-terminal peptides of bovine pancreatic ribonuclease A, including residues 1-13, 1-15 and 1-20 (95, 153, 154); calcitonin M (473); acyclic and cyclic forms of oxytocin (114); and the aromatic region of hen egg-white lysozyme (10, 473).

Table 3. Carbon-13 protonation shifts for amino acids and peptides

Amino acid carbons[b]	Dissociation shift (ppm)[a]		Other
	$-NH_3^+ \rightarrow -NH_2$	$-COOH \rightarrow COO^-$	
carbonyl	-8.6 ± 0.2	-2.6 ± 0.3	
αC	-1.2 ± 0.3	-1.9 ± 0.2	
βC	-4.7 ± 0.4	$-0.8_5 \pm 0.1_5$	
C-2 (His)	+0.3	+0.3	-2.3 ($\geqslant \overset{+}{N}H \rightarrow \geqslant NH$)
C-4 (His)	-1.6	-1.1	-4.7 ($\geqslant \overset{+}{N}H \rightarrow \geqslant NH$)
C-5 (His)	-1.8	+0.6	+1 ($\geqslant \overset{+}{N}H \rightarrow \geqslant NH$)
Terminal peptide residues[c]			
carbonyl	-6 to -9	-3 to -4	
αC	-1 to -2	-2 to -3	
βC	-2 to -4	-1 to -1.5	
γC	0 to -1	-0.3 to -0.4	
Side-chain carbons in peptides[c]			
δC (Lys)[d]	-5	-	
εC (Lys)[d]	-1.1	-	
δC (Glu)	-	∿-5 (carbonyl)	
γC (Glu)	-	∿-4	
γC (Asp)	-	-3 to -4 (carbonyl)	
βC (Cys)	-	-	-1.5 ($-SH \rightarrow -S^-$)
[αC (Cys)]	-	-	-2.6 ($-SH \rightarrow -S^-$)

[a] +, upfield; -, downfield.

[b] Data are from (357), and the data for the carbonyl, αC and βC represent the means for 11 amino acids.

[c] Data are from (89, 153, 172, 222).

[d] Data are for Lys as an N-terminal residue.

C. Use of Paramagnetic Shift and Relaxation Enhancement (Broadening) Probes

The use of lanthanide β-diketonates as NMR shift reagents to resolve the NMR spectra of organic molecules is well documented (90, 291, 370, 372b, 411). Similarly, the hydrated trivalent lanthanide cations have been extensively employed as shift and line-broadening agents in aqueous media to study molecules of biological interest (130, 163, 325, 371, 372). A few details of the principles involved in these phenomena are reproduced here as a prerequisite to an understanding of their application to peptide sequencing.

1. Paramagnetic Shift Principles

For the purpose of illustration let us assume (but see 253-255) that a paramagnetic metal ion M attaches itself to the carboxylate function of a terminal glycine unit as indicated in Figure 9. The induced change in the chemical shift in ppm, Δ, of a nucleus like H_i in Figure 9 due to the presence of M could arise in two ways. These mechanisms are summarized in Eqs. (1) and (2). Full discussion of these relationships may be found elsewhere (33, 130, 163, 254, 325, 370, 372b).

$$\Delta = x\Delta_c + x\Delta_p \quad (1)$$

$$\Delta = xA_iB/T + xCG_i(\theta_i)/r_i^3 \quad (2)$$

The symbols in these expressions are defined as follows: Δ_c, contact shift for the NMR-observed nucleus in the metal complex, which arises from a through-bond coupling interaction with M; Δ_p, pseudocontact shift for the NMR-observed nucleus in the metal complex, which arises from a through-space electron dipole-nuclear dipole interaction; x, fraction of ligand (e.g., peptide) bound to M under conditions of rapid exchange on the NMR time scale; A_i, contact interaction constant for the ith nucleus, which falls off sharply in magnitude with the distance r_i of the ith nucleus from the binding site of M; B, a constant representing the magnetic properties of both M and the nucleus observed (e.g., ^{13}C, ^{1}H); T, absolute temperature; C, a constant representing the magnetic properties (magnitude and symmetry) of M in the complex; $G_i(\theta_i)$, a geometric function of the angle θ_i defined in Figure 9, which reduces to $3\cos^2\theta_i - 1$ for a complex of actual, or what is usually observed in solution, "effective" axial symmetry.

The predominant contribution to Δ appears to be Δ_p for most trivalent lanthanide ions (e.g., Eu^{3+}, Dy^{3+}, Pr^{3+}, Nd^{3+}, Tm^{3+}) for all nuclei removed by one amino acid residue from the metal binding site (46). The matter is not quite as clear for transition metal ion shift probes such as Ni^{2+} and Co^{2+} since Δ_c is known to make substantial contributions to induced shifts in the cases examined [(175), Chap. 4].

It is readily seen from Eq. (2), that in the absence of a contact interaction, Δ contains geometric ($G_i(\theta_i)$) and distance (r_i) information about the complex. Lanthanide-induced shifts, corrected in the few cases where necessary for contact contributions, have been applied in resolving the solution geometries or conformations of both small (30, 130, 163, 253-255, 325) and large (77, 78, 325) biomolecules. In studies with peptides (random coil) and other essentially linear molecules, the geometric term is less important than the r_i^{3-} dependence (253-255). Thus, to a first approximation, the magnitude of the shift is a measure of the proximity of the perturbed nucleus to the paramagnetic shift probe.

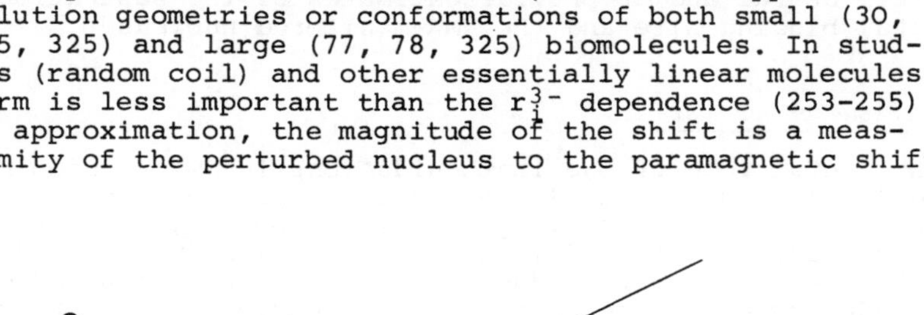

Fig. 9. Definition of the angle θ_i and the distance vector r_i for the binding of the paramagnetic metal ion, M, to the terminal glycine unit of a peptide (see Eq. (2)). Charges are omitted for convenience

2. *Paramagnetic Line Broadening Principles*

The increase in line width ($\Delta\Delta\nu_{1/2}$) due to the presence of a paramagnetic broadening probe, such as the lanthanide ion Gd^{3+} and the transition metal ion Mn^{2+}, in a molecule like that depicted in Figure 9 is expressed quantitatively, but in a simplified manner, in Eqs. (3) and (4). Details may be sought elsewhere (30, 130, 144, 163, 254, 325, 370, 372b).

$$\Delta\Delta\nu_{1/2} = x(\Delta\nu_{1/2}^{con}) + x(\Delta\nu_{1/2}^{dip}) \quad (3)$$

$$\Delta\Delta\nu_{1/2} = xDA_i^2\ f(\omega_e,\tau_e) + x\ E\ f(\omega,\tau_c)/r_i^6 \quad (4)$$

In these expressions x, A_i and r_i have the previously assigned meaning, with the following definitions for the remaining symbols: $\Delta\nu_{1/2}^{con}$ and $\Delta\nu_{1/2}^{dip}$ represent the contact (through-bond) and pseudocontact (through-space) contributions, respectively; D contains the square of the total electron spin angular momentum of M; $f(\omega_e,\tau_e)$ is a function of the electron Larmer frequency ω_e, and τ_e the correlation time modulating the scalar or contact interaction; E contains the product of the squares of the magnetic moments of M and the observed nucleus and may include a parameter correcting for the anisotropy of the metal complex; $f(\omega,\tau_c)$ is a function of the nuclear resonance frequency ω and the correlation time τ_c modulating the dipolar interaction.

In our present survey we will mainly be interested in Gd^{3+} and some of its derivatives, and Cu^{2+} as broadening agents. The contact or scalar term in Eq. (4) is not important for Gd^{3+}, and thus an increase in line width is directly dependent on the sixth power of the reciprocal distance between the metal and the ith nucleus (130, 325). In contrast to the lanthanide shifts, broadening data are not encumbered by a geometric angular dependence, and are thus more reliable in measuring the separation between the metal and the observed nucleus. This aspect makes Gd^{3+} a very useful entity in conformational (vide supra) and primary sequence studies (vide infra).

The Cu^{2+} ion is somewhat more difficult to use as a relaxation enhancement agent, since both the contact and dipolar terms contribute to the observed broadening (144). However, since the hyperfine constant A_i also falls off with distance, selective broadening experiments are still useful and do provide estimates of the separation between the metal binding site and the NMR-monitored nucleus.

3. *Sequence Analysis*

The useful role of the lanthanide ions in NMR sequencing is illustrated in Figures 10 and 11. The partial spectra shown of L-Ala-Gly-Gly and Gly-Gly-L-Ala are very similar at pD 3.9, and were recorded on a 60 mHz spectrometer of moderate resolution (Varian Model T-60). When a little Dy^{3+} ion was added, immediate separation and differential broadening of the resonance peaks occurred. It is obvious that the broadening and displacement is most severe for protons on the C-terminal residue and the effect falls off for those nuclei of the internal and N-terminal residues. If a little Gd^{3+} were now added, those peaks shifting most strongly would also broaden to the greatest extent. This is illustrated for Gly-Gly in Figure 11. Consequently, under the experimental conditions, the lanthanide ions bind to the C-terminus (cf. Fig. 9). Detailed conformational analysis based on lanthanide-induced shift and relaxation enhancement data has shown that peptides

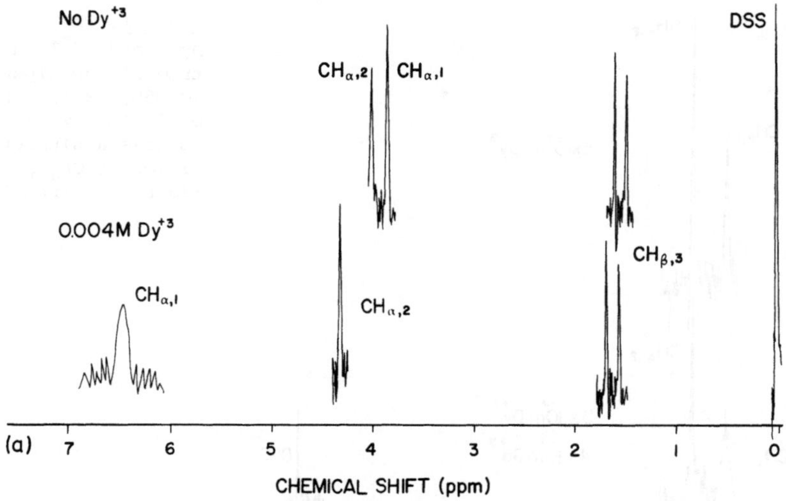

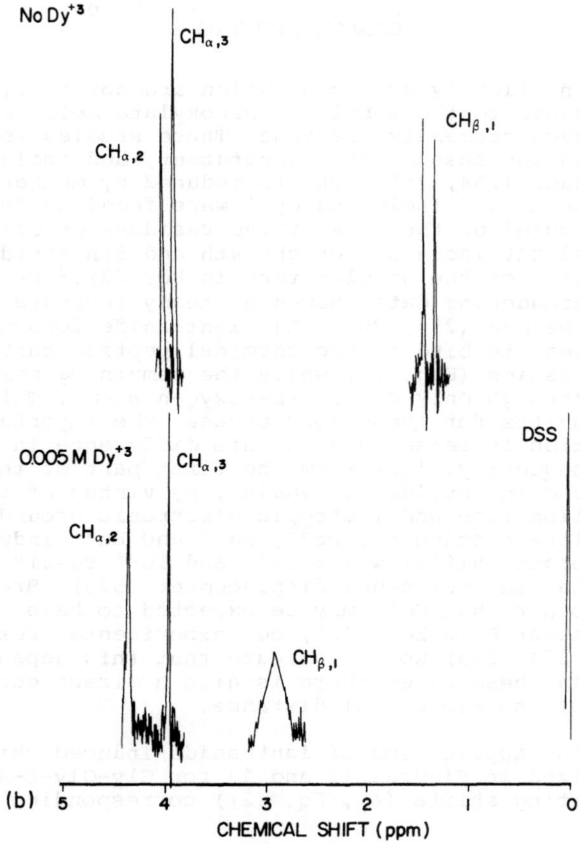

Fig. 10a and b. The proton spectra at 60 MHz of 0.05M solutions (D_2O, pH 3.9) of (a) L-Ala-Gly-Gly and (b) Gly-Gly-L-Ala in the absence or presence of Dy^{3+}. The multiplet resonance of the alanyl methine proton is not shown. $CH_{\alpha,x}$ and $CH_{\beta,x}$ identify the position of a ^1H nucleus in a given residue (α or β) and the residue number (x) from the C-terminus. From Podolski (345)

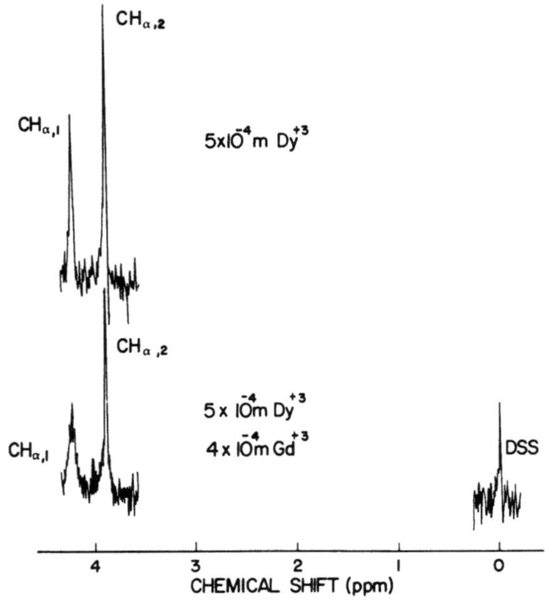

Fig. 11. The combined effect of Dy^{3+} and Gd^{3+} on the proton spectrum of the dipeptide Gly-Gly (0.05M, pH 4). In the absence of Dy^{3+}, the two α-CH_2 peaks overlap to give a singlet. For the definition of $CH_{\alpha,x}$ see legend to Figure 10. From Podolski (345)

in slightly acidic solution are not fully extended, although the distance to the terminal carboxylate metal binding site increases for each successive residue. These studies were carried out in solution in the absence of a denaturant, and included peptides up to five residues (254, 255). Shifts induced by members of the lanthanide series up to and including Dy^{3+} were found to decrease monotonically for ^1H nuclei of the first three residues of oligoglycines, and showed a slight increase for the 4th and 5th residue because of the contribution of the angular term in Eq. (2).[3] Estimated distances from Gd^{3+} broadening data showed a steady increase for each succeeding glycine residue (254, 253). All lanthanide ions up to and including Dy^{3+} appear to bind to the terminal peptide carboxylate moiety in a bidentate fashion (Fig. 9), while the remaining members of the series attach through only one of the oxygen atoms. This complicates the observed shifts for these ions because the importance of the angular contribution increases due to this difference in binding mode (254, 255). Consequently, ions form the first part of the series seem best suited for use in peptide sequencing. By virtue of its long electron-spin relaxation time and isotropic electronic ground state, Gd^{3+} only induces line broadening; Pr^{3+}, Nd^{3+} and Eu^{3+} induce little broadening and moderate shifts; while Dy^{3+} and Tb^{3+} result in both appreciable broadening and resonance displacement (325). Broadening by lanthanide ions other than Gd^{+3} may be expected to have some angular dependence (constant E in Eq. (4)), but experimental results for Dy^{3+} (345) and Tm^{3+} (254, 255) would indicate that this dependence is negligible, and thus in these cases there is also a direct correlation with the sixth power of the reciprocal distance.

The application of lanthanide-induced shifts in sequencing is summarized in Figures 12 and 13 for Gly-Gly-L-Ala and L-Ala-L-Ala-Gly. Limiting shifts (Δ_p, Eq. (2)) corresponding to those for protons in the

---

[3]This reversal was not observed by Podolski when the medium was 3M $NaClO_4$ (vide infra).

Fig. 12. Plot of the reciprocal of the Dy^{3+}-induced shifts ($1/\Delta$) vs. the peptide:metal ratio for the tripeptide Gly-Gly-L-Ala. For the definition of the symbols see legend to Figure 10. From Podolski (345)

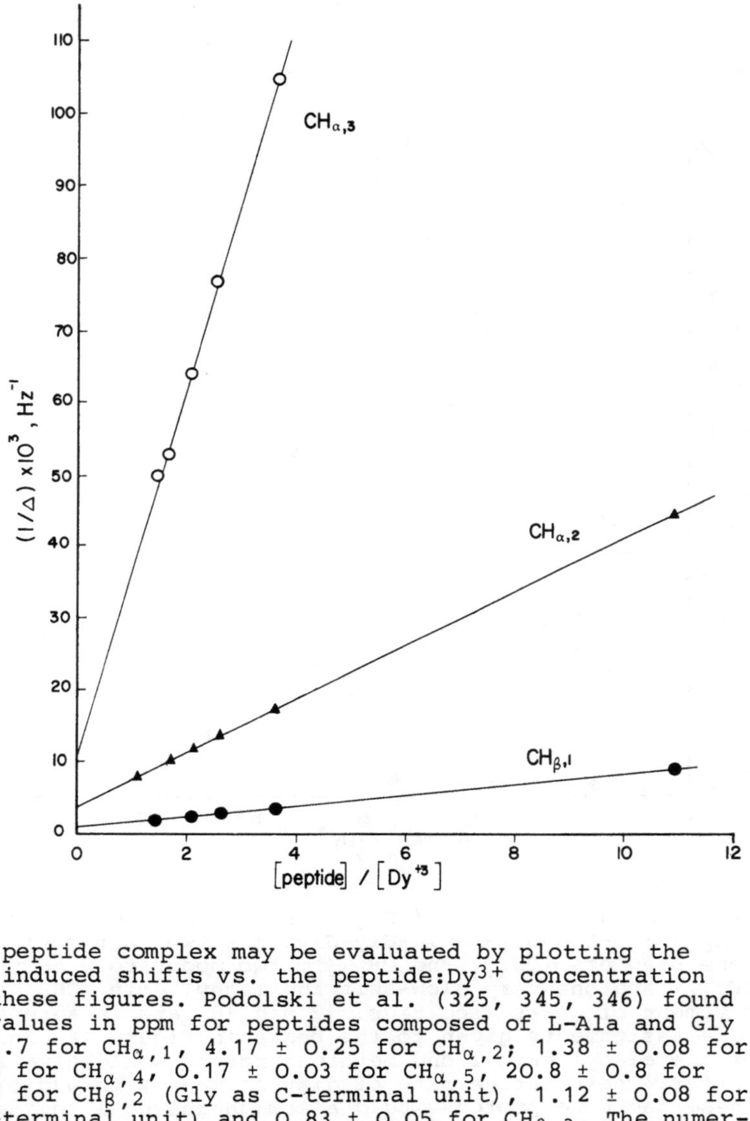

fully formed Dy^{3+} peptide complex may be evaluated by plotting the reciprocal of the induced shifts vs. the peptide:Dy^{3+} concentration ratio as done in these figures. Podolski et al. (325, 345, 346) found the following Δ_p values in ppm for peptides composed of L-Ala and Gly residues: 33.3 ± 1.7 for $CH_{\alpha,1}$, 4.17 ± 0.25 for $CH_{\alpha,2}$; 1.38 ± 0.08 for $CH_{\alpha,3}$, 0.55 ± 0.05 for $CH_{\alpha,4}$, 0.17 ± 0.03 for $CH_{\alpha,5}$, 20.8 ± 0.8 for $CH_{\beta,1}$, 1.67 ± 0.08 for $CH_{\beta,2}$ (Gly as C-terminal unit), 1.12 ± 0.08 for $CH_{\beta,2}$ (L-Ala as C-terminal unit) and 0.83 ± 0.05 for $CH_{\beta,3}$. The numerical subscript refers to the residue number relative to the C-terminus. As indicated earlier, a reversal of the αCH protons of the 4th and 5th residues as reported by Levine et al. (254, 255) was not observed. This may be due to different experimental conditions. To enhance the solubility of pentaglycine, Podolski (345) used 3M $NaClO_4$.

An analysis of the spectra of two synthetic pentapeptides is depicted in Figure 14. In each case, the spectrum in the absence of Dy^{3+} shows three overlapping glycine methylene singlets, and two alanyl methyl doublets (alanyl methine protons are not shown as they are more difficult to resolve due to their multiplet structure). Addition of Dy^{3+} in several stages establishes the identity of each peptide. From the data in Figure 14 and spectra for subsequent Dy^{3+} additions, the following conclusions could be drawn: Gly-L-Ala-Gly-L-Ala-Gly has two

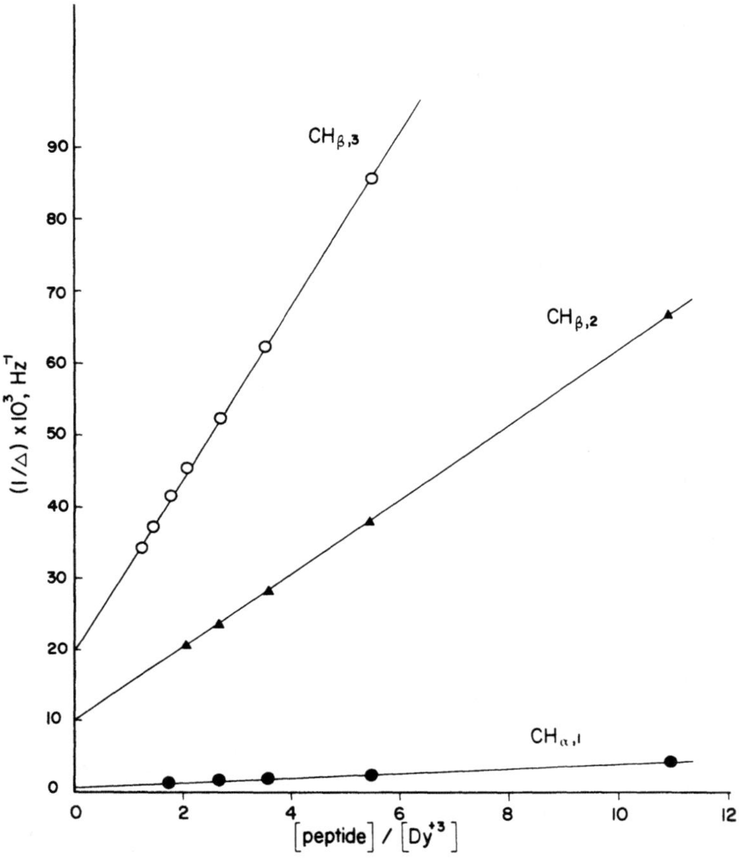

Fig. 13. Plot of the reciprocal of the Dy^{3+}-induced shifts ($1/\Delta$) vs. the peptide-metal ratio for the tripeptide L-Ala-L-Ala-Gly. For the definition of the symbols see legend to Figure 10. From Podolski (345)

fast-moving resonances ($CH_{\alpha,3}$, $CH_{\beta,4}$) and one slow-moving resonance ($CH_{\alpha,5}$); L-Ala-Gly-L-Ala-Gly-Gly, has two fast-moving glycine methylene singlets ($CH_{\alpha,1}$ and $CH_{\alpha,2}$), two moderately shifted peaks ($CH_{\beta,3}$, $CH_{\alpha,4}$) and one very slow-shifting methyl unit ($CH_{\beta,5}$). The rate of movement of peaks (see Figs. 12 and 13) are in agreement with the residue assignments given to the peaks in Figure 14. Selective decoupling of the alanyl methyl and methine protons would give singlets for the latter. The relative rate of displacement with addition of Dy^{3+} of these backbone protons would allow unequivocal residue number assignments. Addition of a little Gd^{3+} to a well-separated spectrum (e.g., by Eu^{3+}, Pr^{3+} or Dy^{3+}) would also clearly identify the residue positions.

A comparison of Cu^{2+} with Gd^{3+} as relaxation enhancement probes is given in Figure 15 for triglycine. It is seen that the protons of the N-terminal residue experience a greater effect than the C-terminal ones when Cu^{2+} is used. Copper ions coordinate preferentially with the amino group in basic solution, while when this function is protonated in acidic solution, Gd^{3+} coordinates exclusively with the carboxylate group. These trends are consistent with the principles of coordination chemistry (324). In basic solution, the tripositive lanthanide ions would precipitate as hydroxide complexes (325, 348). The data in Figure 15 clearly indicate that Gd^{3+} is a more efficient broadening agent

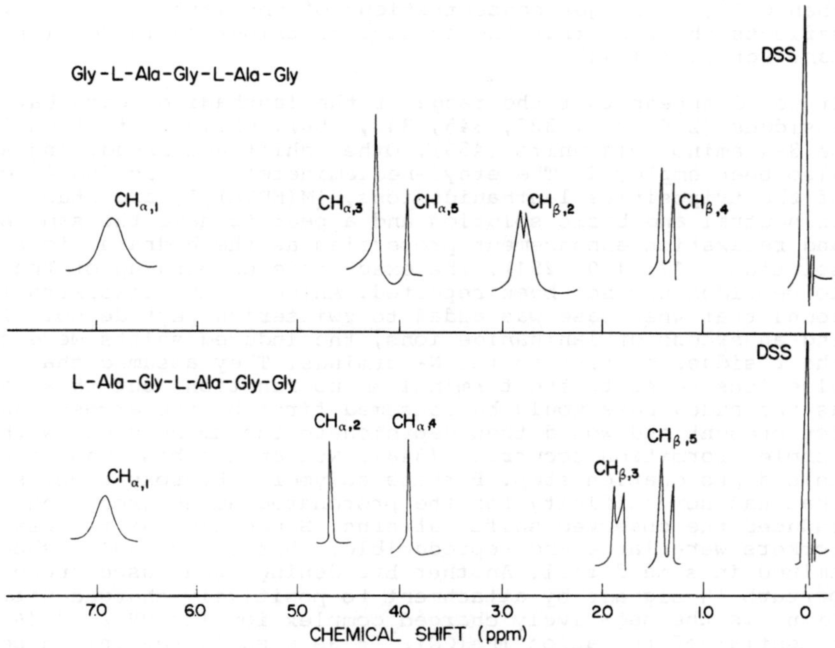

Fig. 14. The proton spectra at 60 MHz of two pentapeptides (∿0.05M, pH ∿4) of identical amino acid composition in the presence of a small amount of Dy^{3+} (0.005M). See text for further explanation, and the legend to Figure 10 for definition of $CH_{\alpha,x}$ and $CH_{\beta,x}$

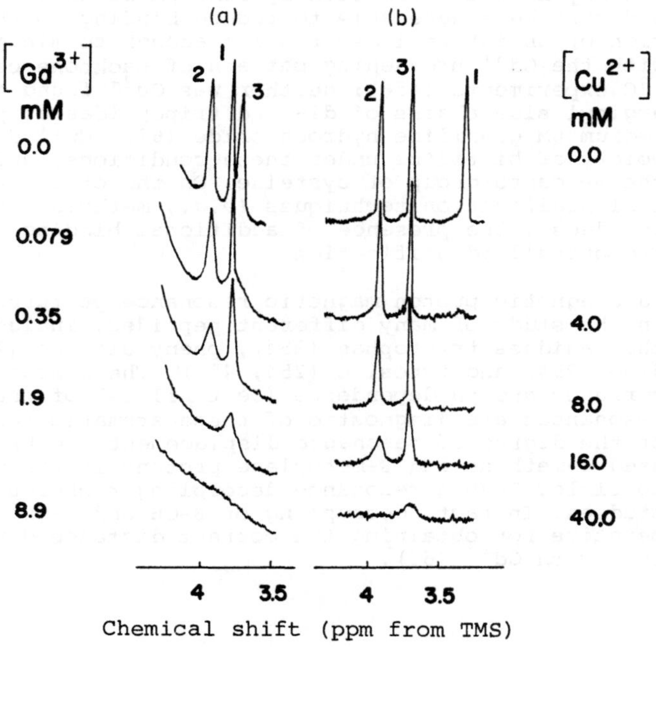

Fig. 15. The proton spectra at 100 MHz of the methylene protons of triglycine numbered sequentially from the N-terminal end: a 2% solution of triglycine in 8M urea at pH 4.3 on addition of Gd^{3+} at the molar concentrations shown; b 4% solution of triglycine in D_2O at pH 11, using Cu^{2+}. Redrawn from Bradbury et al. (63) with permission

than Cu^{2+}, as larger concentrations of the latter are required. This reflects the fact that the contact contribution in Eq. (4) is important for Cu^{2+} (144).

It would appear that the range of the lanthanide perturbations is 4-5 residues (254, 255, 325, 345, 346, 456), while that of Cu^{2+} appears to be 3-4 amino acid units (456). Other shift and broadening agents have also been employed. The ethylenediaminetetraacetic (EDTA) complexes of the tripositive lanthanide ions, $[M(EDTA)]^-$, are stable and soluble in neutral and basic solution and appear to have the same broadening and relaxation enhancement properties as the hydrated ions in acidic solution (117, 139, 251). The exact mode of binding of these complexes to peptides has not been reported. Anteunis and coworkers (18, 19) found that when base was added to zwitterion peptide solutions containing an excess of lanthanide ions, the induced shifts were greatest for the residues nearest to the N-terminus. They assumed that the lanthanide ions bound to the terminal amino function, but this is unlikely as the added base would be consumed first by the excess lanthanide ion present and would then deprotonate the amino group without metal complex formation occurring (348). Not enough base was added to effect this deprotonation step. Perhaps polymeric hydroxy species were formed that had some affinity for the protonated amino group and in this way induced the observed shift patterns. Since the shifts observed by these workers were large and reproducible, this basic medium should be examined in some detail. Another broadening agent used successfully to perturb ^1H signals by attachment to positively charged nitrogen functions is the negatively charged complex ion $[Cr(CN)_6]^{-3}$ (64, 358). The potential of the anion $[Fe(CN)_6]^{-3}$ as a shift reagent in peptide work should also be examined (139, 358). No doubt other suitable probes will also be developed (e.g., 446).

The binding of the perturbing ions to side-chain functional groups is a potential problem. However as mentioned earlier, side-chain amino and carboxylate groups have pK_a values sufficiently different for binding may be made selective by fine pH adjustment. For example, Bradbury and coworkers were able to reduce binding to the carboxylic acid function of an internal Asp residue enough to minimize serious interference with the Gd^{3+} broadening pattern of backbone carbonyl carbon atoms in ^{13}C experiments (456); neither was Cu^{2+} found to bind to the lysyl and arginyl side chains of di- and tripeptides at pH 8 in the denaturing medium 6M guanidine hydrochloride (63). Cu^{2+} does bind to the imidazole moiety of histidine under these conditions, and should also bind to the mercapto group of cysteine. On the other hand, combined with chemical modification techniques (e.g., methylation, protection of terminal residues), the presence of additional binding sites could well help in the overall identification.

Paramagnetic proton magnetic resonance perturbations have been used in the study of many different peptides, including those containing the residues tryptophan (254), phenylalanine (254, 255, 456), histidine (254) and tyrosine (254, 456). The characteristic downfield occurrence and pH dependence (Sect. III.A) of the corresponding proton resonances are diagnostic of these aromatic residues. The examination of the degree of resonance displacement and broadening of the associated α-methine and β-methylene protons in the usual manner is still possible. Double resonance decoupling techniques are useful in such studies. In fact, decoupling of β-CH and α-CH decoupling appears imperative for obtaining the correct distance-dependent broadening pattern with Gd^{+3} (63).

The use of the paramagnetic probes in ^{13}C NMR studies of peptides has also received some attention. Characteristically, the α-carbonyl carbon resonance of the C-terminal residue experiences shifts in the opposite direction from the backbone and β-carbons, and are smaller than those observed for the αC methylene or methine carbons (35). In an Asp (internal) containing pentapeptide, the carbonyl carbons were well separated and could be selectively broadened by Gd^{3+} (vide supra), 456). The remaining carbon atoms could also be broadened selectively. Proton relaxation enhancement was found to be ∿12 times more sensitive than that of the corresponding ^{13}C nucleus (456). Unfortunately, the higher concentration requirement for ^{13}C studies negates the increased resolution attainable with this nucleus.

Finally, protected peptides can also be subject to paramagnetic probing. In polar media, such as methanol or d_6-DMSO, solutions of the anhydrous salts of the metal ions may be used. In aprotic solvents, the β-diketonate derivatives of the lanthanide shift and broadening agents may be employed (370, 411). Association with the peptide unit appears to be through the peptide carbonyl functions (e.g., 234, 282). Nonlanthanide relaxation complexes are also available (258).

D. Sample Preparation

No elaborate sample preparation is required for the NMR analysis of peptides. The peptide is normally dissolved in D_2O, which in the case of paramagnetic perturbation work often contains a denaturant to reduce higher order solution structure. These denaturants also improve the solubility of the longer oligopeptides (e.g., 8M urea (63, 64), 6M guanidine hydrochloride (456), 3M $NaClO_4$ (345, 346)). The pH (or pD) of the peptide solution is then adjusted to the required value with DCl or NaOD. A pH of approximately 4 is required for the lanthanide studies and a pH of 8 for work with Cu^{2+} (63, 64, 345, 346, 456). The solution is then lyophilized several times from 99% or better D_2O, and then finally from 100.0% D_2O. This procedure reduces the HDO peak to an acceptable level, although repeated exposure of the sample to the atmosphere, such as in a titration procedure, will cause it to increase with time. Where necessary, the residual interference of the HDO peak may be removed by double resonance techniques (50, 63, 456).

Most of the data reported in this article are for rather concentrated peptide solutions (0.05-0.10 M). However, the amount of peptide required for sequencing has been reduced to about 0.25 mg (for ^1H studies with a tripeptide at 100 MHz) by use of a micro-tube of 0.05 ml capacity and a small amount of spectral accumulation (63, 64). Use of increased accumulation such as that afforded by FT spectrometers should reduce this to 0.04 mg (456). Future instrumental developments will likely lower the sample requirements even more. Since ^{13}C NMR is intrinsically much less sensitive than ^1H studies, the peptide needed in this case is of the order of 20 mg under the same ideal conditions. Thus proton NMR has a clear advantage here.

The lanthanide ion solutions are quite often prepared from the anhydrous chlorides. However, these salts, when not stored properly, will pick up moisture and CO_2 and will hydrolyze to form rather insoluble complexes. In some cases like Gd^{3+}, soluble hydrated nitrates are available. The lanthanide oxides (e.g., Eu_2O_3) after firing at 900°C for several hours are good primary standards. These may be dissolved in a slight excess of acid (e.g., $Eu_2O_3 + 6H^+ \rightarrow 2Eu^{3+} + 3H_2O$) to give the appropriate stock solutions. For example, the oxide is readily dissolved in D_2O by adding the appropriate amount of DCl and applying gentle heat. A filtering step is often required prior to dilution to remove some residual suspension. Lanthanide solutions must be kept slightly acidic to prevent hydrolysis, since the rare earth hydroxides are only very slightly soluble (~1μm). This problem is especially acute in concentrated solutions (e.g., 10^{-1} M). Stock solutions can be safely stored at pH 3 or less. Since very small aliquots of the lanthanide ions need to be added to the sample (μl quantities), the presence of a little acid does not alter the pH of the peptide solution. Neither is the presence of some HDO serious, since this is unavoidable when the oxides or hydrated nitrates are the starting materials. Aliquots of the metal solution are usually added with a micrometer syringe directly into the NMR tube. For additional details of the preparation and standardization of lanthanide solutions, the reader is referred to the summary of the analytical chemistry of the trivalent rare earth ions given elsewhere (325).

Anhydrous copper(II) chloride should serve as an adequate source for the preparation of stock copper solutions.

IV. Conclusions

We have shown that NMR spectroscopy provides an alternative methodology for peptide sequence analysis. It is possible to sequence peptides of up to 9 or 10 residues by examining the ^{13}C and ^{1}H peptide signals, their pH dependence, their observed spin-spin coupling patterns and chemical shifts, and the perturbations introduced by paramagnetic relaxation and shift probes. Simulation of peptide spectra using compilations of chemical shifts and coupling constants is possible, and might be used to confirm established sequences. The detailed spectral analyses made possible by resolution enhancement techniques afforded by recent Fourier transform technology (77-79) should extend sequencing to considerably longer peptides. Associated with these resolution improvements is the capability of subtracting spectra (77-79). Thus substraction of a perturbed (paramagnetic shifting and broadening; pH shifts) spectrum from an unperturbed one should identify the affected resonances (77-79).

The technology of peptide sequencing by NMR has not yet reached a stage where one would recommend its adoption as a routine procedure for peptides isolated from proteins. The amount of sample required for an NMR analysis is larger than that for chemical methods such as the Edman degradation or for mass spectrometry. For the analysis of synthetic peptides, where the quantity of sample required is not usually a serious obstacle, NMR should be considered as a viable alternative to existing procedures. For the identification of naturally occurring peptides, which frequently show unusual structural features (see Sect. III.A), NMR may provide a new and much needed source of information to complement data obtained by other chemical and physical techniques such as mass spectrometry. Future developments in NMR instrumentation may well reduce the sample requirements to a range acceptable for routine sequencing. An extremely attractive feature of the NMR procedure is its nondestructive nature. Cu^{2+} should be quantitatively removed by precipitation with H_2S followed by centrifugation or filtering. The bulk of a lanthanide ion can be removed by adjusting the pH of the peptide solution to 8-9 and separating the insoluble hydroxide (64b). Any residual lanthanide may be extracted with ethyl acetate following complex formation with 8-hydroxyquinoline (64b). We have found that precipitation of the lanthanide ion with ammonium carbonate, followed by filtration or centrifugation also affects good separation. The excess precipitating agent may simply be removed by lyophilization.

Addendum: Bradbury and Warren (64b) have recently reported the sequence determination of submilligram amounts of peptides by measuring spin-lattice relaxation times, T_1, of α-C protons. The percentage reduction in T_1 values was greatest for the C-terminus which bound the added Gd^{3+}, and decreased sequentially along the chain. Fourier transform NMR facilities are required for the 180°-τ-90° pulse technique used to measure T_1 values (149, 212, 473).

References

1. Abderhalden, E., Brockmann, H.: Biochem. Z. 225, 386 (1930).
2. Adams, J.M.: J. Mol. Biol. 33, 571 (1968).
3. Adams, M.J., Blundell, T.L., Dodson, E.J., Dodson, G.G., Vijayan, M. Baker, E.N., Harding, M.M., Hodgkin, D.C., Rimmer, B., Sheat, S.: Nature (London) 224, 491 (1969).
4. Adams, M.J., Ford, G.C., Kolkoek, R., Lentz, Jr., P.J., McPherson, Jr., A., Rossmann, M.G., Smiley, I.E., Schevitz, R.W., Wonacott, A.J.: Nature (London) 227, 1098 (1970).
5. Adams, M.J., Ford, G.C., Lentz, Jr., P.J., Liljas, A., Rossmann, M.G.: Proc. Nat. Acad. Sci. 70, 1793 (1973).
6. Adman, E.T., Sieker, L.C., Jensen, L.H.: J. Biol. Chem. 248, 3987 (1973).
7. Agrawal, K.L., Johnstone, R.A.W., Kenner, G.W., Millington, D.S., Sheppard, R.C.: Nature (London) 219, 498 (1968).
8. Akabori, S., Ohno, K., Narita, K.: Bull. Chem. Soc. Japan 25, 214 (1952).
9. Alden, R.A., Birktoft, J.J., Kraut, J., Robertus, J.D., Wright, C.S.: Biochem. Biophys. Res. Commun. 45, 337 (1971).
10. Allerhand, A., Childers, R.F., Oldfield, E.: Biochemistry 12, 1335 (1973).
11. Ambler, R.P.: Biochem. J. 96, 32P (1965).
12. Ambler, R.P.: In: Chemotaxonomy and Serotaxonomy. Hawkes, J.G. (ed.). London: Academic Press 1968, p. 57.
13. Anastasi, A., Erspamer, V.: Experientia 18, 58 (1962).
14. Andersen, R.D., Apgar, P.A., Burnett, R.M., Darling, G.D., Lequesne M.E., Mayhew, S.G., Ludwig, M.L.: Proc. Nat. Acad. Sci. 69, 3189 (1972).
15. Andersson, C.O.: Acta Chem. Scand. 12, 1353 (1958).
16. Anderson, J.C., Barton, M.A., Gregory, R.A., Hardy, P.M., Kenner, G.W., McLeod, J.K., Preston, J., Sheppard, R.C., Morley, J.S.: Nature (London) 204, 933 (1964).
17. Andersson, B., Nyman, P.O., Strid, L.: Biochem. Biophys. Res. Commun. 48, 670 (1972).
18. Anteunis, M., Gelan, J.: J. Am. Chem. Soc. 95, 6502 (1973).
19. Anteunis, M., Callens, R.: J. Magn. Resonance 15, 317 (1974).
20. Armentrout, R.W.: Biochim. Biophys. Acta 191, 756 (1969).
21. Armentrout, R.W., Doolittle, R.F.: Arch. Biochem. Biophys. 132, 80 (1969).
22. Arnone, A., Bier, C.J., Cotton, F.A., Day, V.W., Hazen, E.E., Jr., Richardson, D.C., Richardson, J.S., Jonath, A.: J. Biol. Chem. 246, 2302 (1971).
23. Arpino, P.J., McLafferty, F.W.: Determination of Organic Structure by Physical Methods. New York: Academic Press 1975, Vol. VI, p. 10[illegible]
24. Atassi, M.Z., Tarlowski, D.P., Paull, J.H.: Biochim. Biophys. Acta 221, 623 (1970).
25. Atassi, M.Z., Habeeb, A.F.S.: Methods Enzymol. 25, 546 (1972).
26. Atassi, M.Z.: In: Specific Receptors of Antibodies, Antigens, and Cells. Basel: Karger 1973, p. 118.
27. Augusteyn, R.C., Spector, A.: Biochem. J. 124, 345 (1971).

28. Bacon, V., Jellum, E., Patton, W., Pereira, W., Halpera, B.: Biochem. Biophys. Res. Commun. 37, No. 6, 878 (1969).
29. Baldwin, M.A., McLafferty, F.W.: Org. Mass Spec. 7, 1111 (1973).
30. Banner, D.W., Bloomer, A.C., Petsko, G.A., Phillips, D.C., Pogson, C.I., Wilson, I.A., Corran, P.H., Furth, A.J., Milman, I.D., Offord, R.E., Priddle, J.D., Waley, S.G.: Nature (London) 255, 609 (1975).
31. Barber, M., Powers, P., Wallington, M.J., Wolstenholme, W.A.: Nature (London) 212, 784 (1966).
32. Barrett, G.C.: Chem. Commun. 487 (1967).
33. Barry, C.D., Glasel, J.A., Williams, R.J.P., Xavier, A.V.: J. Mol. Biol. 84, 471 (1974).
34. Batchelor, J.G., Feeney, J., Roberts, G.C.K.: J. Magn. Resonance 20, 19 (1975).
35. Bayer, E., Beyer, K.: Tetrahedron Lett. 15, 1209 (1973).
36. Beecham, A.F.: Tetrahedron Lett. 3, 211 (1967).
37. Beecham, A.F., Ham, N.S.: Tetrahedron Lett. 24, 2773 (1968).
38. Benjamini, E., Scibienski, R.J., Thompson, K.: Contemp. Topics Immunochem. 1, 1 (1972).
39. Bernal, J.D., Crowfoot, D.C.: Nature (London) 133, 794 (1934).
40. Berson, S.A., Yalow, R.S.: J. Clin. Invest. 38, 2019 (1959).
41. Beuhler, R.J., Flanigan, E., Greene, L.J., Friedman, L.: Biochemistry 13, 5060 (1974).
42. Beynon, J.H.: Mass Spectrometry and its Applications to Organic Chemistry. Amsterdam: Elsevier 1960.
43. Beynon, J.H., Saunders, R.A., Williams, A.E.: The Mass Spectra of Organic Molecules. Amsterdam: Elsevier 1968.
44. Biemann, K., Gapp, F., Seibl, J.: J. Am. Chem. Soc. 81, 2274 (1959).
45. Biemann, K.: In: Mass Spectrometry: Organic Chemical Applications. New York: McGraw-Hill 1962.
46. Bieman, K., Cone, C., Webster, B.R., Arsenault, G.P.: J. Am. Chem. Soc. 88, 5598 (1966).
47. Blake, C.C.F., Koenig, D.F., Mair, G.A., North, A.C.T., Phillips, D.C., Sarma, V.R.: Nature (London) 206, 757 (1965).
48. Blake, C.C.F., Evans, P.R.: J. Mol. Biol. 84, 585 (1974).
49. Blake, C.C.F., Geisow, M.J., Swan, I.D.A., Rerat, B., Rerat, C.: J. Mol. Biol. 88, 1 (1974).
50. Bleich, H.E., Glasel, J.A.: J. Magn. Resonance 18, 401 (1975).
51. Blombäck, D.: Methods Enzymol. 11, 398 (1968).
52. Blombäck, B., Doolittle, R.F.: Acta Chem. Scand. 17, 1816 (1963).
53. Blombäck, B., Doolittle, R.F.: Acta Chem. Scand. 17, 1819 (1963).
54. Blout, E.R., DesRoches, M.E.: J. Am. Chem. Soc. 81, 370 (1959).
55. Blout, E.R., Idelson, M.: J. Am. Chem. Soc. 80, 4909 (1958).
56. Blout, E.R.: In: Fundamental Aspects and Recent Developments in Optical Rotatory Dispersion and Circular Dichroism, Proc. NATU Advan. Study Inst. Ciardelli, F., Salvadori, P. (eds). London: Heyden 1971, p. 352-372.
57. Blow, D.M., Janin, J., Sweet, R.M.: Nature (London) 249, 54 (1974).
58. Bode, W., Schwager, P.: FEBS Lett. 56, 139 (1975).
59. Boissonas, R.A., Sandrin, E.: Experientia 18, 59 (1962).
60. Bovey, F.A.: Nuclear Magnetic Resonance Spectroscopy. New York: Academic Press 1969.
61. Bovey, F.A.: In: High Resolution NMR of Macromolecules. New York: Academic Press 1972.
62. Bradshaw, R.A., Erickson, L.H., Walsh, K.A., Neurath, H.: Proc. Nat. Acad. Sci. 63, 1389 (1969).
63. Bradbury, J.H., Brown, L.R., Crompton, M.W., Warren, B.: Pure Appl. Chem. 40, 83 (1974).
64. Bradbury, J.H., Crompton, M.W., Warren, B.: Anal. Biochem. 62, 310 (1974).
65. Bradshaw, R.A., Ericsson, L.H., Neurath, H., Walsh, K.A.: Proc. Nat. Acad. Sci. 63, 1389 (1969).
66. Bratton, A.C., Marshall, E.K.: J. Biol. Chem. 128, 537 (1939).

67. Braunitzer, G., Matsuda, G.: J. Biol. Chem. 53, 262 (1963).
68. Bricas, E., van Heijenoort, J., Barber, M., Wolstenholme, W.A., Das, B.C., Lederer, E.: Biochemistry 4, 2254 (1965).
69. Bridgen, J.: FEBS Lett. 50, 159 (1975).
70. Bridgen, J.: In: Solid-Phase Methods in Protein Sequence Analysis. Rocsford, Ill.: Pierce Chemical Co. 1975, p. 11.
71. Buchner, M., Ford, G.C., Moras, D., Olsen, K.W., Rossmann, M.G.: J. Mol. Biol. 90, 25 (1974).
72. Budzikiewicz, H., Djerassi, C., Williams, D.H.: In: Interpretation of Mass Spectra of Organic Compounds. San Francisco: Holden-Day 1964.
73. Bundi, A., Grathwohl, C., Hochmann, J., Keller, R.M., Wagner, G., Wuetrich, K.: J. Magn. Resonance 18, 191 (1975).
74. Buse, G., Braig, S., Braunitzer, G.: Hoppe-Seylers. Z. Physiol. Chem. 349, 1686 (1969).
75. Butler, P.J.G., Hartley, B.S.: Methods Enzymol. 25, 191 (1972).
76. Cambell, J.W., Watson, H.C., Hodgson, G.I.: Nature (London) 250, 301 (1974).
77. Campbell, I.D., Dobson, C.M., Williams, R.J.P.: Proc. Roy. Soc. (London) Ser. A. 345, 23 (1975).
78. Campbell, I.D., Dobson, C.M., Williams, R.J.P.: Proc. Roy. Soc. London Ser. A. 345, 41 (1975).
79. Campbell, I.D., Xavier, A.V.: J. Magn. Resonance 11, 172 (1973).
80. Campbell, I.D., Xavier, A.V., Jeminet, G., Williams, R.J.P.: FEBS Lett. 49, 115 (1974).
81. Canfield, R.E.: J. Biol. Chem. 238, 2698 (1963); 240, 1997 (1965).
82. Cann, J.R., Stewart, J.M., Matsueda, G.R.: Biochemistry 12, 3780 (1973).
83. Capra, J.D., Kehoe, J.M.: Advan. Immunol. 20, 1 (1975).
84. Caprioli, R.M., Seifert, W.E.: Biochem. Biophys. Res. Commun. 64, 295 (1975).
85. Carter, C.W., Freer, S.T., Xuong, Ng.H., Alden, R.A., Kraut, J.: Cold Spring Harbor Symp. Quant. Biol. 36, 381 (1971).
86. Celada, F., Strom, R.: Quart. Rev. Biophys. 5, 395 (1972).
87. Chang, J.K., Sievertsson, H., Bogentoft, C., Currie, B., Folkers, K., Daves, G.D., Jr.: J. Med. Chem. 14, 481 (1971).
88. Chen, R., Doolittle, R.F.: Biochemistry 10, 4486 (1971).
89. Christl, M., Roberts, J.D.: J. Am. Chem. Soc. 94, 4565 (1972).
90. Cockerill, A.F., Davies, G.L.O., Harden, R.C., Rackham, D.M.: Chem. Rev. 73, 553 (1973).
91. Cocks, G.T., Wilson, A.C.: Science 164, 188 (1969).
92. Cohen, J.S., Jardetzky, O.: Proc. Nat. Acad. Sci. 60, 92 (1968).
93. Cohen, J.S., Feil, M., Chaiken, I.M.: Biochim. Biophys. Acta 236, 468 (1971).
94. Cohen, J.S.: Biochim. Biophys. Acta 229, 603 (1971).
95. Cohen, J.S.: Exper. Methods Biophys. Chem. 6, 521 (1972).
96. Cohn, M.: Quart. Rev. Biophys. 3, 61 (1970).
97. Colman, P.M., Jansonius, J.N., Matthews, B.W.: J. Mol. Biol. 70, 701 (1972).
98. Cotton, A.: Compt. Rend. 120, 989 (1859).
99. Craig, L.C.: Proc. Nat. Acad. Sci. 61, 152 (1968).
100. Crawford, J.L., Lipscomb, W.N., Schellman, C.G.: Proc. Nat. Acad. Sci. 70, 538 (1973).
101. Crumpton, M.J.: Biochem. J. 100, 223 (1966).
102. Cunningham, B.A., Wang, J.L., Waxdal, M.J., Edelman, G.M.: J. Biol Chem. 250, 1503 (1975).
103. Das, B.C., Gero, S.D., Lederer, E.: Biochem. Biophys. Res. Commun. 29, 211 (1967).
104. Das, B.C., Lederer, E.: In: New Techniques in Amino Acid, Peptide and Protein Analysis. Niederwieser, A., Pataki, G. (eds.). Ann Arbor: Science Publishers 1971.

105. Davidson, B.E., Sajgó, M., Noller, H.G., Harris, J.I.: Nature (London) 216,1181 (1967).
106. Day, R.A., Falter, H., Lehman, J.P., Hamilton, R.E.: J. Org. Chem. 38, 782 (1973).
107. Day, R.A., Evans, J.V., Jayasimhulu, K.: Abstr. 28, Div. Biol. Chem. 170th Am. Chem. Soc. Mtg. Chicago 1975.
108. Dayhoff, M.O.: Atlas of Protein Sequence and Structure. Silver Spring, Md.: Natl. Biomed. Res. Found. 1972, Vol. V.
109. Dayhoff, M.O.: Atlas of Protein Sequence and Structure. Silver Spring, Md.: Natl. Biomed. Res. Found. 1973, Vol. V, Supp. I.
110. Deisenhoffer, J., Steigeman, W.: Acta Cryst. B 31, 238 (1975).
111. DeHaas, G.H., Franck, F., Keil, B., Thomas, D.W., Lederer, E.: FEBS Lett. 4, 25 (1969).
112. Dekker, C.A., Stone, D., Fruton, J.S.: J. Biol. Chem. 181, 719 (1949).
113. Dell, A., Morris, H.R.: Biochem. Biophys. Res. Commun. 61, 1125 (1974).
114. Deslauriers, R., Walter, R., Smith, I.C.P.: Biochem. Biophys. Res. Commun. 48, 854 (1972).
115. Dickerson, R.E., Takano, T., Eisenberg, D., Kallai, O.B., Samson, L., Cooper, A., Margoliash, E.: J. Biol. Chem. 246, 1511 (1971).
116. Dickerson, R.F., Timkovich, R., Almassy, R.J.: J. Mol. Biol. 100, 473 (1976).
117. Dobson, C.M., Williams, R.J.P., Xavier, A.V.: J. Chem. Soc. 1762, 1765 (1974).
118. Doolittle, L.R., Mross, G.A., Fothergill, L.A., Doolittle, R.F.: Anal. Biochem. 78, 491 (1977).
119. Dowling, L.M.: Methods Enzymol. 25B, 231 (1972).
120. Dowling, L.M., Armentrout, R.W.: Abstracts, 7th Intern. Congr. Biochem. Tokyo, 1967.
121. Dowling, L.M., Armentrout, R.W.: Biochemistry 7, 516 (1968).
122. Dowling, L.M., Cottrell, B.A., Riley, M.: In press.
123. Dowling, L.M.: Methods Enzymol. 25, 231 (1972).
124. Dowling, L.M., Stark, G.R.: Biochemistry 8, 4728 (1969).
125. Drenth, J., Jansonius, J.N., Koekoek, R., Swen, H.M., Wolters, B.G.: Nature (London) 218, 929 (1968).
126. Drenth, J., Hol, W.G., Jansonius, J.N., Koekoek, R.: Cold Spring Harbor Symp. Quant. Biol. 36, 107 (1971).
127. Drude, P.: In: Lehrbuch der Optik, 2nd ed. Leipzig: Hirzel 1906.
128. Dus, K., Sletton, K., Kamen, M.D.: J. Biol. Chem. 243, 5507 (1968).
129. Dus, K., Tedro, S., Bartsch, R.G., Kamen, M.D.: Biochem. Biophys. Res. Commun. 43, 1239 (1971).
130. Dwek, R.A.: In: Nuclear Magnetic Resonance (NMR) in Biochemistry. Oxford: Clarendon Press 1973.
131. Edelman, G.M., Cunningham, B.A., Reeke Jr., G.N., Becker, J.W., Waxdal, M.J., Wang, J.L.: Proc. Nat. Acad. Sci. 69, 2580 (1972).
132. Edman, P.: Arch. Biochem. 22, 475 (1949).
133. Edman, P.: In: Protein Sequence Determination. Needleman, S.B. (ed.). Berlin-Heidelberg-New York: Springer 1970, p. 212.
134. Edman, P., Begg, G.: Europ. J. Biochem. 1, 80 (1967).
135. Edmundson, A.B., Hirs, C.H.W.: Nature (London) 190, 663 (1961).
136. Edmundson, A.B.: Nature (London) 205, 883 (1965).
137. Eisenberg, D.: The Enzymes. 1 (3rd ed.), 1 (1970).
138. Eklund, H., Nordström, B., Zeppezauer, E., Söderlund, G., Ohlsson, I., Boiwe, T., Brändén, C.I.: FEBS Lett. 44, 200 (1974).
139. Elgavish, G.A., Reuben, J.: J. Am. Chem. Soc. 98, 4755 (1976).
140. Elion, J., Labie, D.: In: Solid-Phase Methods in Protein Sequence Analysis. Rockford, Ill.: Pierce Chemical Co. 1975, p. 73.
141. Emsley, J.W., Feeney, J., Sutcliffe, L.H.: High Resolution in Nuclear Magnetic Resonance Spectroscopy. Vols. I and II. Oxford: Pergamon Press 1965.

142. Epp, O., Colman, P., Fehlhammer, H., Bode, W., Schiffer, M., Huber, R., Palm, W.: Europ. J. Biochem. 45, 513 (1974).
143. Ernst, R.R.: J. Chem. Phys. 45, 3845 (1966).
144. Espersen, W.G., Martin, R.B.: J. Am. Chem. Soc. 98, 40 (1976).
145. Fales, H.M.: In: Mass Spectrometry: Techniques and Applications. Milne, G.W.A. (ed.). New York: Wiley-Interscience 1971.
145b. Falter, H.: Ph.D. Dissertation, University of Cincinnati, 1968.
146. Falter, H., Jayasimhulu, K., Day, R.A.: Analyt. Biochem. 67, 359 (1975).
147. Falter, H., Waern, I.: Abstract PH-99, 59th Can. Chem. Conf. London, Ontario, June 1976.
148. Farr, R.S.: J. Infect. Dis. 103, 239 (1958).
149. Farrar, T.C., Becker, E.D.: In: Pulse and Fourier transform NMR. New York-London: Academic Press 1971.
150. Fermandjian, S., Morgat, J.L., Fromageot, P.: Europ. J. Biochem. 24, 252 (1971).
151. Fernlund, P., Josefsson, L.: Science 177, 173 (1972).
152. Flikweert, J.P., Heerma, W., Penders, T.J., Dijkstra, G., Arens, J.F.: Recueil 86, 293 (1967).
153. Freedman, M.H., Lyerla, J.R., Chaiken, I.M., Cohen, J.S.: Europ. J. Biochem. 32, 215 (1973).
154. Freedman, M.H., Cohen J.S., Chaiken, I.M.: Biochem. Biophys. Res. Commun. 42, 1148 (1971).
155. Freer, S.T., Alden, R.A., Carter, C.W., Kraut, J.: J. Biol. Chem. 250, 46 (1975).
156. Freer, S.T., Kraut, J., Robertus, J.D., Wright, H.T., Xuong, Ng.H. Biochemistry 9, 1997 (1970).
157. Freudenberg, K., Meister, M.: Ann. 518, 86 (1935).
158. Fuchs, S., Cautrecasas, P., Ontjes, D.A., Anfinsen, C.B.: J. Biol. Chem. 244, 943 (1964).
159. Fujiki, H., Braunitzer, G., Rudloff, V.: Z. Physiol. Chemie 351, 901 (1970).
160. Furie, B., Schechter, A.N., Sachs, D.H., Anfinsen, C.B.: Biochemistry 13, 1561 (1974).
161. Furth, A.J., Milman, J.D., Priddle, J.D., Offord, R.E.: Biochem. J. 139, 11 (1974).
162. Gilpin, J.A.: Anal. Chem. 31, 935 (1959).
163. Glasel, J.A.: Progr. Inorg. Chem. 18, 383 (1973).
164. Grathwohl, C., Wuethrich, K.: J. Magn. Resonance 13, 217 (1974).
165. Gray, W.R.: Methods Enzymol. 25, 121 (1972).
166. Greenstein, J.P., Winitz, M.: In: Chemistry of the Amino Acids. New York: Wiley 1961, Vol. I.
167. Greenstein, J.P., Winitz, M.: In: Chemistry of the Amino Acids. New York: Wiley 1961, Vol. II, pp. 920, 1246.
168. Gregory, H., Hardy, P.M., Jones, D.S., Kenner, G.W., Sheppard, R.: Nature (London) 204, 931 (1964).
169. Griffen, J.H., Furie, B., Schechter, A.N.: Biochimie 57, 453 (1975).
170. Gross, E., Witkop, B.: J. Biol. Chem. 237, 1856 (1962).
171. Gross, E.: Methods Enzymol. 11, 238 (1967).
172. Gurd, F.R.N., Lawson, P.J., Cochran, D.W., Wenkert, E.: J. Biol. Chem. 246, 3725 (1971).
173. Hakomori, S.I.: J. Biochem. 55, 205 (1964).
174. Hamilton, R.E., Patil, G.V., Jayasimhulu, K., Day, R.A.: Org. Mass Spec. 9, 211 (1974).
175. Hamming, M.C., Foster, N.G.: In: Interpretation of Mass Spectra of Organic Compounds. New York: Academic Press 1972.
176. Haralambidou, E., Day, R.A.: Org. Mass Spec. 10, 683 (1975).
177. Hardman, K.D., Ainsworth, C.F.: Biochemistry 11, 4910 (1972).
178. Hartley, B.S.: Nature (London) 201, 1284 (1964).
179. Hartley, B.S.: Phil. Trans. Roy. Soc. London B 257, 77 (1970).

180. Hass, G.M., Nau, H., Biemann, K., Grahn, D.T., Ericsson, L.H., Neurath, H.: Biochemistry 14, 6, 1334 (1975).
181. Heidelberger, M., Kendall, F.E.: J. Exp. Med. 62, 697 (1935).
182. Heil, A., Müller, G., Noda, L.H., Pinder, T., Schirmer, I., Schirmer, R.H., von Zahern, I.: Europ. J. Biochem. 43, 131 (1974).
183. Heller, W.J.: Phys. Chem. 62, 1569 (1958).
184. Henderson, L.E., Henriksson, D., Nyman, P.O., Strid, L.: Proc. Alfred Benzon Symp. 4, 341 (1971).
185. Henderson, L.E., Henriksson, D., Nyman, P.O.: Biochem. Biophys. Res. Commun. 52, 1388 (1973).
186. Herriot, J.R., Watenpangh, K.D., Sieker, L.C., Jensen, L.H.: J. Mol. Biol. 80, 423 (1973).
187. Herriot, J.R., Sieker, L.C., Jensen, L.H., Lovenberg, W.: J. Mol. Biol. 50, 391 (1970).
188. Herskowitz, T.T.: J. Am. Chem. Soc. 86, 4445 (1964).
189. Heyns, K., Grutzmacher, H.F.: Tetrahedron Lett. 26, 1761 (1963).
190. Hill, E., Tsernaglou, D., Webb, L., Banaszak, L.J.: J. Mol. Biol. 72, 577 (1972).
191. Hirs, C.H.W., Moore, S., Stein, W.H.: J. Biol. Chem. 219, 623 (1956).
192. Hoare, D.G., Koshland, D.E., Jr.: J. Biol. Chem. 242, 2447 (1967).
193. Holmes, J.L., Benoit, F.M.: In: Mass Spectrometry, MTP International Review of Science. Maccoll, A. (ed.). London: Butterworths Univ. Park Press 1972.
194. Holmgren, A.: Europ. J. Biochem. 6, 475 (1968).
195. Holmgren, A., Söderberg, B.O., Eklund, H., Brändén, C.I.: Proc. Nat. Acad. Sci. 72, 2305 (1975).
196. Hood, L., Gray, W.R., Dreyer, W.J.: J. Mol. Biol. 22, 179 (1966).
197. Horn, M.J., Laursen, R.A.: FEBS Lett. 36, 285 (1973).
198. Horn, M.J.: Anal. Biochem. In press.
199. Horsley, W.J., Sternlicht, H.: J. Am. Chem. Soc. 90, 3738 (1968).
200. Horsley, W.J., Sternlicht, H., Cohen, J.S.: Biochem. Biophys. Res. Commun. 37, 47 (1969).
201. Horsley, W.J., Sternlicht, H., Cohen, J.S.: J. Am. Chem. Soc. 92, 680 (1970).
202. Huber, R., Kukla, D., Bode, W., Schwager, P., Bartels, K., Deisenhoffer, J., Steigeman, W.: J. Mol. Biol. 89, 73 (1974).
203. Huber, R., Steigemann, W.: FEBS Lett. 48, 235 (1974).
204. Huber, R., Kukla, D., Rühlmann, A., Epp, O., Formanek, H.: Naturwissenschaften 57, 389 (1970).
205. Huber, R., Epp, O., Steigemann, W., Formanek, H.: Europ. J. Biochem. 19, 42 (1971).
206. Hudson, G., Biemann, K.: Biochem. Biophys. Res. Commun. 71, 212 (1976).
207. Husain, S.S., Lowe, G.: Biochem. J. 116, 689 (1970).
208. Iizuka, E., Yang, J.T.: Biochemistry 3, 1519 (1964).
209. Ikenaka, T., Bammerlin, H., Kaufman, H., Schmid, K.: J. Biol. Chem. 241, 5560 (1966).
210. Imamura, T., Baldwin, T.O., Riggs, A.: J. Biol. Chem. 247, 2785 (1972).
211. Jackman, L.M., Sternhell, S.: In: Applications of Nuclear Magnetic Resonance Spectroscopy in Organic Chemistry, 2nd ed. Elmsford, N.Y.: Pergamon 1969.
212. James, T.L.: In: Nuclear Magnetic Resonance in Biochemistry. New York: Academic Press 1975.
213. Javid, J., Pettis, P.K.: J. Lab. Clin. Med. 86, 777 (1975).
214. Jayasimhulu, K., Day, R.A.: Anal. Biochem. 2, 359 (1975).
215. Jennings, K.R.: In: Mass Spectrometry: Techniques and Applications. Milne, G.W.A. (ed.). New York: Wiley-Interscience 1971.
216. Jensen, K.A., Pedersen, C.: Acta Chem. Scand. 15, 1087 (1961).
217. Jirgensons, B.: Optical Rotatory Dispersion of Proteins and Other Macromolecules. In: Molecular Biology, Biochemistry and Biophysics, Vol. 5. Berlin-Heidelberg-New York: Springer 1969.

218. Jirgensons, B.: Biochem. Biophys. Acta 434, 58 (1976).
219. Jollès, J., Jauregui-Adell, J., Jollès, P.: Biochim. Biophys. Acta 78, 68 (1963).
220. Jörnvall, H.: Europ. J. Biochem. 16, 25 (1970).
221. Svec, H.J., Junc, G.A.: J. Am. Chem. Soc. 86, 2278 (1964).
222. Jung, G., Breitmaier, E., Voelter, W.: Europ. J. Biochem. 24, 438 (1972).
223. Kabat, E.A., Mayer, M.M.: Experimental Immunochemistry, 2nd ed. Springfield: Thomas 1971, p. 241.
224. Kannan, K.K., Liljas, A., Waara, I.: Bergstén, P.C., Lövgren, S., Strandberg, B., Bengtsson, U., Carlbom, U., Fridborg, K., Järup, L., Petef, M.: Cold Spring Harbor Symp. Quant. Biol. 34, 221 (1975).
225. Strandberg, B., Notstrand, B., Fridborg, K., Lövgren, S., Ohlsson, A., Petef, M.: The Stockholm Symposium on The Structure of Biological Molecules. Abstracts, 76 pp. (1973).
226. Strandberg, B., Fridborg, K., Lövgren, S., Notstrand, B., Ohlsson, A., Petef, M.: Proc. Nat. Acad. Sci. 72, 51 (1975).
227. Kartha, G., Bello, J., Harker, D.: Nature (London) 213, 862 (1967)
228. Kassell, B., Laskowski, M.S.: Biochem. Biophys. Res. Commun. 20, 463 (1965).
229. Kawasaki, I., Itano, H.A.: Anal. Biochem. 48, 546 (1972).
230. Kay, E.: J. Biol. Chem. 251, 3411 (1976).
231. Kendrew, J.C., Dickerson, R.E., Strandberg, B.E., Hart, R.G., Davies, D.R., Phillips, D.C., Shore, V.C.: Nature (London) 185, 422 (1960).
232. Kendrew, J.C., Watson, H.C., Strandberg, B.E., Dickerson, R.E., Phillips, D.C., Shore, V.C.: Nature (London) 190, 666 (1961).
233. Kenner, G.W., Khorana, H.G.: J. Chem. Soc. (London) 2, 2076 (1952)
234. Kessler, H., Molter, M.: Abstract A-7, 6th International Conference on Magnetic Resonance in Biological Systems, Kandersteg, Switzerland, Sept. 16-21, 1974.
235. Kiser, R.W.: Introduction to Mass Spectrometry and its Applications. Englewood Cliffs, N.J.: Prentice-Hall 1965.
236. Kjaer, A.: Acta Chem. Scand. 4, 1347 (1950).
237. Koide, T., Tsunasawa, S., Ikenaka, T.: J. Biochem. 71, 165 (1972).
238. Kolb, B.: In: New Techniques in Amino Acid, Peptide and Protein Analysis. Niederwieser, A., Pataki, G. (eds.). Ann Arbor: Science Publ. 1971.
239. Kreil, G., Kreil-Kiss, G.: Biochem. Biophys. Res. Commun. 27, 275 (1967).
240. Kuhn, R., Löw, I., Trischmann, H.: Chem. Ber. 90, 203 (1957).
241. Kunkel, H.G., Agnello, V., Joslin, F.G., Winchester, R.J., Capra, J.D.: J. Exp. Med. 137, 331 (1973).
242. Langer, B., Steinmetz-Kayne, M., Hilschmann, N.: Hoppe-Seylers, Z. Physiol. Chem. 349, 945 (1968).
243. Lasker, S.E. (ed.): Proc. Intern. Conf. on Electron Spin Resonance and Nuclear Magnetic Resonance in Biology and Medicine and Magnetic Resonance in Biological Systems. Ann. N.Y. Acad. Sci. 222, 1-1124 (1973).
244. Laursen, R.A.: J. Am. Chem. Soc. 88, 5344 (1966).
245. Laursen, R.A.: In: Immobilized Enzymes, Antigens, Antibodies and Peptides. Weetall, H.H. (ed.). New York: Marcel Dekker 1975, p. 56
246. Laursen, R.A., Bonner, A.G., Horn, M.H.: In: Instrumentation in Amino Acid Sequence Analysis. Perham, R.N. (ed.). New York: Academic Press 1976.
247. Laursen, R.A.: Solid-Phase Methods in Protein Sequence Analysis. Rockford, Ill.: Pierce Chemical Co. 1975.
248. Laursen, R.A., Horn, M.J., Bonner, A.G.: FEBS Lett. 21, 67 (1972).
249. Laursen, R.A.: Europ. J. Biochem. 20, 89 (1971).
250. Laursen, R.A.: Methods Enzymol. 25B, 344 (1972).

251. Lee, C., Raszka, M.J.: J. Magn. Resonance 17, 151 (1975).
252. Lenard, J., Gallop, P.M.: Anal. Biochem. 29, 203 (1969).
253. Levine, B.A., Thornton, J.M., Williams, R.J.P.: J.C.S. Chem. Commun. 669 (1974)
254. Levine, B.A., Williams, R.J.P.: Proc. R. Soc. London, Ser. A. 345, 5 (1975).
255. Levine, B.A., Williams, R.J.P.: In: Environmental Effects on Molecular Structure and Properties. Pullman, B. (ed.), p. 95. Dordrecht-Holland: D. Reidel Publishing Co. 1976.
256. Levsen, K., Wipf, H.K., McLafferty, F.W.: Org. Mass Spectrom. 8, 117 (1974).
257. Levy, G.C., Nelson, G.L.: Carbon-13 Nuclear Magnetic Resonance for Organic Chemists. New York: Wiley-Interscience 1972.
258. Levy, G.C., Komoroski, R.A.: J. Am. Chem. Soc. 96, 678 (1974).
259. Li, S.L., Riggs, A.: J. Biol. Chem. 245, 6149 (1970).
260. Light, A., Frater, R., Kimmel, J.R., Smith, E.L.: Proc. Nat. Acad. Sci. 52, 1276 (1964).
261. Liljas, A., Kannan, K.K., Bergsten, P.C., Waara, I., Fridborg, K., Strandberg, B., Carlbom, U., Järup, L., Lövgren, S., Petef, M.: Nature New Biol. 235, 131 (1972).
262. Lin, K.D., Deutsch, H.F.: J. Biol. Chem. 248, 1885 (1973).
262a. Lin, K.D., Deutsch, H.F.: J. Biol. Chem. 249, 2329 (1974).
263. Ling, N., Rivier, J., Burgus, R., Guillemin, R.: Biochemistry 12, 5307 (1973).
264. Lipscomb, W.N., Hartsuck, J.A., Reeke Jr., G.N., Quiocho, F.A., Bethge, P.H., Ludwig, M.L., Steitz, T.A., Muirhead, H., Coppola, J.C.: Structure Function and Evolution in Proteins. Brookhaven Symp. Biol. 21, 24 (1968).
265. Lipscomb, W.N., Hartsuck, J.A., Quiocho, F.A., Reeke, G.N. Jr.: Proc. Nat. Acad. Sci. 64, 28 (1969).
266. Liu, T.Y.: Methods Enzymol. 25, 44 (1972).
267. Love, W.E., Klock, P.A., Lattman, E.E., Padlan, E.A., Ward Jr., K.B., Hendrickson, W.A.: Cold Spring Harbor Symp. Quant. Biol. 36, 349 (1971).
268. Lowry, T.M.: In: Optical Rotatory Power. London: Longmans Green 1935.
269. Ludoweig, J., Dorfman, A.: Biochim. Biophys. Acta 38, 212 (1960).
270. Lutz, O., Jirgensons, B.: Ber. deut. chem. Ges. 63, 448 (1930).
271. Lutz, O., Jirgensons, B.: Ber. deut. chem. Ges. 64, 1221 (1931).
272. Macfarlane, R.D., Torgerson, D.F.: Science 191, 920 (1976).
273. Machleidt, W., Hofner, H., Wachter, E.: In: Solid-Phase Methods in Protein Sequence Analysis. Laursen, R.A. (ed.). Rockford, Ill.: Pierce Chemical Co. 1975, p. 17.
274. Mahler, H.R., Cordes, E.H.: Biological Chemistry. New York: Harper and Row 1966.
275. Mandel, M.: J. Biol. Chem. 240, 1586 (1965).
276. Manusadzhyan, V.G., Zyakoon, A.M., Chuvilin, A.V., Varshavskii, Y.M.: Isv. Akad. Nauk. Arm. SSR, Ser. Khim. 17, 143 (1964).
277. Margoliash, E., Smith, E.L., Kreil, G., Tuppy, H.: Nature (London) 192, 1121 (1961).
278. Margoliash, E., Nisonoff, A., Reichlin, M.: J. Biol. Chem. 245, 931 (1970).
279. Markland, F.S., Smith, E.L.: J. Biol. Chem. 242, 5198 (1967).
280. Markley, J.L.: Accounts Chem. Res. 8, 70 (1975).
281. Martin, D.R., Williams, R.J.P.: Biochem. J. 153, 181 (1976).
282. Martinelli, L.C., Honigberg, I.L., Sternson, L.A.: Tetrahedron Lett. 29, 1671 (1973).
283. Marvel, C.S., Radzitzky, P. de, Brader, J.J.: J. Am. Chem. Soc. 77, 5997 (1955).
284. Matthews, B.W.: The Proteins 3 (3rd ed.) in press.

285. Matthews, B.W.: J. Mol. Biol. 33, 491 (1968).
286. Matthews, B.W., Sigler, P.B., Henderson, R., Blow, D.M.: Nature (London) 214, 652 (1967).
287. Matthews, B.W., Argos, P., Levine, M.: Cold Spring Harbor Symp. on Quant. Biol. 36, 387 (1971).
288. Matthews, B.W., Jansonius, J.N., Colman, P.M., Schoenborn, B.P., Dupourque, D.: Nature New Biol. 238, 37 (1972).
289. Matthews, B.W., Colman, P.M., Jansonius, J.N., Titani, K., Walsh, K.A., Neurath, H.: Nature New Biol. 238, 41 (1972).
290. Matthews, B.W., Remington, S.J.: Proc. Nat. Acad. Sci. 71, 4178 (1974).
291. Mayo, B.C.: Chem. Soc. Rev. 2, 49 (1973).
292. McCarthy, K.: Ph.D. Dissertation, George Washington University (1972).
293. McDonald, C.C., Phillips, W.D.: J. Am. Chem. Soc. 91, 1513 (1969).
294. McDonald, C.C., Phillips, W.D.: In: Fine Structure of Proteins and Nucleic Acids. Fasman, G.D., Timasheff, N. (eds.). New York: Dekker 1970, Vol. IV, p. 1.
295. McIver, R.T., Ledford Jr., E.B., Miller, J.S.: Anal. Chem. 47, 692 (1974).
296. McLafferty, F.W.: Analyt. Chem. 31, 82 (1959).
297. McLafferty, F.W.: In: Interpretation of Mass Spectra, 2nd ed. London: Benjamin 1973.
298. Merrifield, R.B.: J. Am. Chem. Soc. 85, 2149 (1963).
299. Mikes, O., Holeysovsky, V., Tomasek, V., Sorm, F.: Biochem. Biophys. Res. Commun. 24, 346 (1966).
300. Mildvan, A.S., Cohn, M.: Adv. Enzymol. 33, 1 (1970).
301. Mitchel, R.E.J., Chaiken, I.M., Smith, E.L.: J. Biol. Chem. 245, 3485 (1970).
302. Mitchell, S.: In: The Cotton Effect. London: Bell 1933.
303. Moews, P.C., Kretsinger, R.H.: J. Mol. Biol. 91, 201 (1975).
304. Morris, H.R., Dickinson, R.J., Williams, D.H.: Biochem. Biophys. Res. Commun. 51, 247 (1973).
305. Morris, H.R., Dudley, W.H., Ambler, R.P.: Biochem. J. 125, 189 (1971).
306. Morris, H.R., Williams, D.H., Midwinter, G.G., Hartley, B.S.: Biochem. J. 141, 701 (1974).
307. Morris, H.R.: FEBS Lett. 22, 257 (1972).
308. Moscowitz, A.: In: Optical Rotatory Dispersion, Application to Organic Chemistry. Djerassi, C. (ed.). New York: McGraw-Hill 1960, p. 150.
309. Mross, G.A., Doolittle, R.F.: Federation Proc. 30, 1241 (1971).
310. Mross, G.A.: Ph.D. Dissertation, Univ. California, San Diego, 1971.
311. Nakamura, A., Jardetzky, O.: Biochemistry 7, 1226 (1968).
312. Nakamura, A., Jardetzky, O.: Proc. Nat. Acad. Sci. 58, 2212 (1967)
313. Nakamura, S., Iwanaga, S., Suzuki, T., Mikuni, Y., Konishi, K.: Biochem. Biophys. Res. Commun. 58, 250 (1974).
314. Nakayama, T., Titani, K., Narita, K.: J. Biochem. 70, 311 (1971).
315. Narita, K.: Biochim. Biophys. Acta 28, 184 (1958).
316. Natvig, J.B., Kunkel, H.G.: Adv. Immunol. 16, 1 (1973).
317. Nau, H.: Angew. Chem. Intern. Ed. Engl. 15, 75 (1976).
318. Nau, H.: Biochem. Biophys. Res. Commun. 59, 1088 (1974).
319. Nau, H., Biemann, K.: Analyt. Biochem. 73, 139 (1976).
320. Nau, H., Biemann, K.: Analyt. Biochem. 73, 154 (1976).
321. Nau, H., Biemann, K.: Analyt. Biochem. 73, 175 (1976).
322. Nau, H., Biemann, K.: Analyt. Chem. 46, 426 (1974).
323. Nau, H., Kelly, J.A., Forster, H.J., Biemann, K.: Biomed. Mass Spectrom. 2, 326 (1975).
324. Nieboer, E., McBryde, W.A.E.: Can. J. Chem. 51, 2512 (1973).
325. Nieboer, E.: Struct. Bonding 22, 1 (1975).
326. Nisonoff, A., Reichlin, M., Margoliash, E.: J. Biol. Chem. 245, 940 (1970).

327. Noble, R.W., Reichlin, M., Gibson, Q.H.: J. Biol. Chem. 244, 2403 (1969).
328. Noble, R.W., Reichlin, M., Schreiber, R.D.: Biochemistry 11, 3326 (1972).
329. Nockolds, C.E., Kretsinger, R.H., Coffee, C.J., Bradshaw, R.A.: Proc. Nat. Acad. Sci. 60, 58 (1972).
330. Notstrand, B., Vaara, I., Kannan, K.K.: Isozymes, Molecular Structure, Proc. 3rd Intern. Symposium on Isozymes. Markert, C. (ed.). New York: Academic Press 1975, p. 575.
331. Okada, K., Nagai, S., Uyehara, T., Hiramoto, M.: Tetrahedron Lett. 30, 1175 (1974).
332. Olaitan, S.A., Delange, R.J., Smith, E.L.: J. Biol. Chem. 243, 5296 (1968).
333. Ovchinnikov, Yu.A., Kiryushkin, A.A., Kozhevnikova, I.V.: Gen. Chem. USSR 41, 2105 (1971).
334. Ozols, J., Strittmatter, P.: J. Biol. Chem. 244, 6617 (1969).
335. Page, M.I., Jencks, W.P.: J. Am. Chem. Soc. 94, 8818 (1972).
336. Patil, G.V., Hamilton, R.E., Day, R.A.: Org. Mass Spec. 7, 817 (1973).
337. Patthy, L., Smith, E.: J. Biol. Chem. 250, 557 (1975).
338. Paul, R., Anderson, G.W.: J. Am. Chem. Soc. 82, 4596 (1960).
339. Payne, J.W., Jakes, R., Hartley, B.S.: Biochem. J. 117, 757 (1970).
340. Pechere, J.F., Capony, J.P., Ryden, L.: Europ. J. Biochem. 23, 421 (1971).
341. Perutz, M.F., Muirhead, H., Cox, J.M., Goaman, L.C.G., Matthews, F.S., McGandy, E.L., Webb, L.E.: Nature (London) 219, 29 (1968).
342. Pfaender, P.: Justus Liebig's Ann. Chem. 707, 209 (1967).
343. Phillips, D.C.: Advan. Struct. Res. Diffraction Methods 2, 75 (1966).
344. Phillips, D.M.P.: Biochem. J. 86, 397 (1962).
345. Podolski, M.L.: The Lanthanide Cations as NMR Probes: Amino Acid Sequence Determination of Simple Peptides and Conformational Studies of the Hormone 3,5,3'-triiodo-L-thyronine. M.Sc. thesis. Laurentian Univ. 1974.
346. Podolski, M.L., Nieboer, E., Falter, H.: In press.
347. Poljak, R.J., Amzel, L.M., Avey, H.P., Chen, B.L., Phizackerley, R.P., Saul, F.: Proc. Nat. Acad. Sci. 10, 3305 (1973).
348. Prados, R., Stadtherr, L.G., Donato, H. Jr., Martin, R.B.: J. Inorg. Nucl. Chem. 36, 689 (1974).
349. Prager, E.M., Fainaru, M., Wilson, A.C., Arnon, R.: Immunochemistry 11, 153 (1974).
350. Prager, E.M., Wilson, A.C.: J. Biol. Chem. 246, 5978 (1971).
351. Prager, E.M., Wilson, A.C.: J. Biol. Chem. 246, 7010 (1971).
352. Press, E.M., Pigott, P.J., Porter, R.R.: Biochem. J. 99, 356 (1966).
353. Previero, A., Pechère, J.F.: Biochem. Biphys. Res. Commun. 40, 549 (1970).
354. Previero, A., Derancourt, J., Coletti-Previero, M.A., Laursen, R. A.: FEBS Lett. 33, 135 (1973).
355. Pritchard, D.G., Schnute Jr., W.C., Todd, C.W.: Biochem. Biophys. Res. Commun. 65, 312 (1975).
356. Prox, A., Sun, K.K.: Z. Naturforsch. 21B, 1028 (1966).
357. Quirt, A.R., Lyerla, J.R., Peat, I.R., Cohen, J.S., Reynolds, W.F., Freedman, M.H.: J. Am. Chem. Soc. 96, 570 (1974).
358. Radda, G.K., Williams, R.J.P.: Chem. Br. 12, 124 (1976).
359. Reeke Jr., G.N., Hartsuck, J.A., Ludwig, M.L., Quiocho, F.A., Steitz, T.A., Piscomb, W.N.: Proc. Nat. Acad. Sci. 58, 2220 (1967).
360. Reich, H.J., Jautelat, M., Messe, M.T., Weigert, F.J., Roberts, J.D.: J. Am. Chem. Soc. 91, 7445 (1969).
361. Reichlin, M., Hay, M., Levine, L.: Biochemistry 2, 971 (1963).
362. Reichlin, M., Bucci, E., Antonini, E., Wyman, J., Rossi-Fanelli, A.: J. Mol. Biol. 9, 785 (1964).

363. Reichlin, M., Fogel, S., Nisonoff, A., Margoliash, E.: J. Biol. Chem. 241, 255 (1966).
364. Reichlin, M.: Immunochemistry 15 (1970).
365. Reichlin, M., Nisonoff, A., Margoliash, E.: J. Biol. Chem. 245, 947 (1970).
366. Reichlin, M.: J. Mol. Biol. 64, 485 (1972).
367. Reichlin, M.: Immunochemistry 11, 21 (1974).
368. Reichlin, M.: Adv. Immunol. 20, 71 (1975).
369. Reindel, F., Hoppe, W.: Chem. Ber. 87, 1103 (1954).
370. Reuben, J.: Progr. NMR Spectrosc. 9, 1 (1973).
371. Reuben, J.: Naturwissenschaften 62, 172 (1975).
372. Reuben, J.: In: Handbook of the Physics and Chemistry of Rare Earths. Gschneider, Jr., K.A., Eyring, L. (eds.), Chap. 39. Amsterdam: North-Holland Publ. Co. 1977.
372b. Reuben J., Elgavish, G.A.: In: Handbook on the Physics and Chemistry of Rare Earths. Gschneider, Jr., K.A., Eyring, L. (eds.), Chap. 38. Amsterdam: North-Holland Publ. Co. 1977.
373. Richards, F.M.: J. Mol. Biol. 37, 225 (1968).
374. Richardson, J.S., Thomas, K.A., Rubin, B.H., Richardson, D.C.: Proc. Nat. Acad. Sci. 72, 1349 (1975).
375. Roberts, G.C.K., Jardetzky, O.: Advan. Protein Chem. 24, 447 (1970).
376. Robinson, C., Bott, M.J.: Nature (London) 168, 325 (1951).
377. Roboz, J.: Introduction to Mass Spectrometry. Instrumentation and Techniques. New York: Wiley-Interscience 1968.
378. Rühlmann, A., Kukla, D., Schwager, P., Bartels, K., Huber, R.: J. Mol. Biol. 77, 417 (1973).
379. Rydon, H.N., Hardy, P.M., Haylock, J.C., Marlborough, D.I., Storey, H.T.: Macromolecules 4, 435 (1971).
380. Ryhage, R.: Anal. Chem. 36, 759 (1964).
381. Ryle, A.P., Sanger, F., Smith, L.F., Kitai, R.: Biochem. J. 60, 541 (1955).
382. Sachs, D.H., Schechter, A.N., Eastlake, A., Anfinsen, C.B.: J. Immunol. 109, 1300 (1972).
383. Sachs, D.H., Schechter, A.N., Eastlake, A., Anfinsen, C.B.: Proc. Nat. Acad. Sci. 69, 3790 (1972).
384. Sachs, D.H., Schechter, A.N., Cohen, J.S.: J. Biol. Chem. 246, 6576 (1971).
385. Sachs, H., Brand, E.: J. Am. Chem. Soc. 76, 1811 (1954).
386. Sakato, N., Fujio, H., Amano, T.: Biken J. 14, 405 (1971).
387. Salemme, F.R., Freer, S.T., Xuong, Ng.H., Alden, R.A., Kraut, J.: J. Biol. Chem. 248, 3910 (1973).
388. Schellenberger, A., Jeschkeit, H., Henkel, R., Lehmann, H.: Z. Chem. 7, 191 (1967).
389. Schellenberger, A., Graubaum, H., Meck, C., Sternkop, G.: Z. Chem. 12, 62 (1972).
390. Schellman, C.G., Schellman, J.A.: Compt. rend. trav. lab. Carlsberg, Ser. Chim. 30 No. 27 (1958).
391. Schellman, J.A.: In: Optical Rotatory Dispersions, Applications to Organic Chemistry. Djerassi, C. (ed.). New York: McGraw-Hill 1960, p. 210-228.
392. Schier, G.M., Halpern, B.: Australian J. Chem. 27, 2455 (1974).
393. Schiffer, M., Girling, R.L., Ely, K.R., Edmundson, A.B.: Biochemistry 12, 4620 (1973).
394. Schiltz, E.: In: Solid-Phase Methods in Protein Sequence Analysis. Rockford, Ill.: Pierce Chemical Co. 1975, p. 47.
395. Schmer, G., Kreil, G.: Anal. Biochem. 29, 186 (1969).
396. Schroer, T., Doolittle, R.F.: Unpublished experiments.
397. Schulz, G.E., Elzinga, M., Marx, F., Schirmer, R.H.: Nature (London) 250, 120 (1974).
398. Segal, D.M., Powers, J.C., Cohen, G.H., Davies, D.R., Wilcox, P.E.: Biochemistry 10, 3728 (1971).

399. Sela, M.: Science 166, 4365 (1969).
400. Senn, M., Venkataraghavan, R., McLafferty, F.W.: J. Am. Chem. Soc. 88, 5593 (1966).
401. Sheinblatt, M.: J. Am. Chem. Soc. 88, 2845 (1966).
402. Sheinblatt, M.: J. Am. Chem. Soc. 87, 572 (1965).
403. Sheinblatt, M.: J. Magn. Resonance 8, 55 (1972).
404. Shemyakin, M.M., Ovchinnikov, Y.A., Kiryushkin, A.A., Vinogradova, E.I., Miroshnikov, A.I., Alakhov, Y.B., Lipkin, V.M., Shetsev, Y.B., Wulfson, N.S., Rosinov, B.V., Bocharev, V.N., Burikov, V.M.: Nature (London) 211, 361 (1966).
405. Shemyakin, M.M., Ovchinnikov, A., Vinogradova, E.I., Feigina, M.Yu., Kiryushkin, A.A., Aldanova, N.A., Alakhov, Yu.B., Lipkin, V.M., Rosinov, B.V.: Experientia 23, 428 (1967).
406. Shemyakin, M.M.: Pure Appl. Chem. 17, 313 (1968).
407. Shemyakin, M.M., Vinogradova, E.I., Ovchinnikov, Yu.A., Kiryushkin, A.A., Feigina, M.Yu., Aldanova, N.A., Alakhov, Yu.B., Lipkin, V.M., Rosinov, B.V., Fonina, L.A.: Tetrahedron Lett. 25, 5785 (1969).
408. Shemyakin, M.M., Ovchinnikov, Yu.A., Kiryushkin, A.A.: In: Mass Spectrometry: Techniques and Applications. Milne, G.W.A. (ed.). New York: Wiley-Interscience 1971.
409. Shotton, D.M., Watson, H.C.: Nature (London) 235, 811 (1970).
410. Shotton, D.M., Hartley, B.S.: Nature (London) 225, 802 (1970).
411. Sievers, R.E.: Nuclear Magnetic Resonance Shift Reagents. New York: Academic Press 1973.
412. Smit, J.D.G.: Thesis Rijksuniversiteit Te Groningen, 1973.
413. Smit, J.D.G., Ploegman, J.H., Kalk, K.H., Jansonius, J.N., Drenth, J.: Stockholm Symp. Structure Biol. Molecules. Abstract, 54 p. (1973).
414. Smith, D.B.: Can. J. Biochem. 42, 755 (1963).
415. Smith, I.: In: Chromatographic and Electrophoretic Techniques. New York: Interscience 1960, Vol. I.
416. Smithies, O., Gibson, D., Fanning, E.M., Goodfliesh, R., Gilman, J.G., Ballantyne, D.L.: Biochemistry 10, 4912 (1971).
417. Smyth, D.G., Stein, W.H., Moore, S.: J. Biol. Chem. 238, 227 (1963).
418. Steinman, H.M., Naik, V.R., Abernethy, J.L., Hill, R.L.: J. Biol. Chem. 249, 7326 (1974).
419. Stenhagen, E., Ryhage, R.: J. Lip. Res. I, 361 (1960).
420. Stothers, J.B.: Carbon-13 NMR Spectroscopy. New York: Academic Press 1972.
421. Stroud, R.M., Kay, L.M., Dickerson, R.E.: J. Mol. Biol. 83, 185 (1974).
422. Szewczuk, A., Kwiatkowska, J.: Europ. J. Biochem. 15, 92 (1970).
423. Szewczuk, A., Mulczyk, M.: Europ. J. Biochem. 8, 63 (1969).
424. Takagi, T., Doolittle, R.F.: Biochemistry 13, 750 (1974).
425. Takahashi, S., Cohen, L.A.: Biochemistry 8, 864 (1969).
426. Takano, T., Swanson, R., Kallai, O.B., Dickerson, R.E.: Cold Spring Harbor Symp. Quant. Biol. 36, 397 (1971).
427. Takano, T., Kallai, O.B., Swanson, R., Dickerson, R.E.: J. Biol. Chem. 248, 5234 (1973).
428. Tanaka, M., Haniu, M., Matsueda, G.: J. Biol. Chem. 246, 3953 (1971).
429. Taniuchi, H., Anfinsen, C.B., Sodja, A.: J. Biol. Chem. 242, 4752 (1967).
430. Taylor, S.S., Oxley, S.S., Allison, W.S., Kaplan, N.O.: Proc. Nat. Acad. Sci. 70, 1790 (1973).
431. Thomas, D.W.: Biochem. Biophys. Res. Commun. 33, 483 (1968).
432. Thomas, D.W., Lederer, E., Bodanszky, M., Izdebski, J., Muramatsu, I.: Nature (London) 220, 580 (1968).
433. Thomas, D.W., Das, B.C., Gero, S.D., Lederer, E.: Biochem. Biophys. Res. Commun. 32, 199 (1968).
434. Tilander, B., Strandberg, B., Fridborg, K.: J. Mol. Biol. 12, 740 (1965).

435. Timkovich, R., Margoliash, E., Dickerson, R.E.: J. Biol. Chem. (in press).
436. Titani, K., Wikler, M., Shinoda, T., Putman, F.W.: J. Biol. Chem. 245, 2171 (1970).
437. Titani, K., Hermodson, M.A., Ericsson, L.H., Walsh, K.A., Neurath, H.: Nature New Biol. 238, 35 (1972).
438. Tsugita, A., Inouye, M.: J. Mol. Biol. 37, 201 (1968).
439. Tsugita, A., Kobyashi, M., Tani, S., Kyo, S., Rashid, M.A., Yoshida, Y., Rajihara, T., Hagihara, B.: Proc. Nat. Acad. Sci. 67, 442 (1970).
440. Tsuneda, J.N., Yasunobu, K.T., Whiteley, H.R.: J. Biol. Chem. 243, 6262 (1968).
441. Tulinsky, A., Vandlen, R.L., Morimoto, C.N., Mani, N.V., Wright, L.H.: Biochemistry 12, 4185 (1973).
442. Tzunazawa, S., Narita, K., Ogata, K.: Proc. Japan Acad. Sci. 46, 960 (1970).
443. Uliana, J.A., Doolittle, R.F.: Arch. Biochem. Biophys. 131, 561 (1969).
444. Urry, D.W.: Research/Develop. 24, 28 (1973).
445. Urry, D.W.: Research/Develop. 25, 20 (1974).
446. Vainer, L.M., Zolin, V.F., Koreneva, L.G., Sagdeev, R.Z.: Koord. Khim 1, 1512 (1975); Chem. Abstr. 84, 6740 8s (1976).
447. Vanaman, T.C., Wakil, S.J., Hill, R.L.: J. Biol. Chem. 243, 6420 (1968).
448. Van Heijenoort, J., Bricas, E., Das, B.C., Lederer, E., Wolstenholme, W.A.: Tetrahedron Lett. 23, 3403 (1967).
449. Van't Klooster, H.A., Vaarkamp-Lijnse, J.S., Dijkstra, G.: Advan. Mass Spectrom. 6, 1027 (1974).
450. Venkatachalam, C.M.: Biopolymers 6, 1425 (1968).
451. Vetter-Diechtl, H., Vetter, W., Richter, W., Biemann, K.: Experientia 24, 341 (1968).
452. Vilkas, E., Lederer, E.: Tetrahedron Lett. 26, 3089 (1968).
453. Wachter, E., Machleidt, W., Hofner, H., Otto, J.: FEBS Lett. 35, 97 (1973).
454. Wachter, E., Hofner, H., Machleidt, W.: In: Solid-Phase Methods in Protein Sequence Analysis. Rockford, Ill.: Pierce Chemical Co. 1975, p. 31.
455. Ward. D.N., Coffey, J.A.: Biochemistry 3, 1575 (1964).
456. Warren, B., Bradbury, J.H.: J. Polymer Sci. 49, 65 (1975).
457. Watenpaugh, K.D., Sieker, L.C., Herriot, J.R., Jensen, L.H.: Cold Spring Harbor Symp. Quant. Biol. 36, 359 (1971).
458. Watenpaugh, K.D., Sieker, L.C., Jensen, L.H., Legall, J., Dubourdien, M.: Proc. Nat. Acad. Sci. 69, 3185 (1972).
459. Watson, H.C., Kendrew, J.C., Coulter, C.L., Brändén, C.I., Phillips, D.C., Blake, C.F.: Acta Cryst. 16 A 84 (1963).
460. Watson, H.C., Shotton, D.M., Cox, J.M., Muirhead, H.: Nature (London) 225, 806 (1970).
461. Watson, J.T., Biemann, K.: Anal. Chem. 37, No. 7, 844 (1965).
462. Weetal, H.H.: Nature (London) 223, 959 (1969).
463. Weigert, F.J., Jautelat, M., Roberts, J.D.: Proc. Nat. Acad. Sci. 60, 1152 (1968).
464. Wenkert, E., Clouse, A.O., Cochran, D.W., Doddrell, D.: J. Am. Chem. Soc. 91, 6879 (1969).
465. Weygand, F., Prox, A., Konig, W., Fessel, H.H.: Angew. Chem. 75, 724 (1963).
466. Weygand, F., Prox, A., Fessel, H.H., Sun, K.K.: Z. Naturforsch. 20b, 1169 (1965).
467. Williams, A.W., Reichlin, M.: Federation Proc. 27, 259 (1968).
468. Winkler, H.U., Beckey, H.D.: Biochem. Biophys. Res. Commun. 46, 391 (1972).
469. Wipf, H.K., Irving, P., McCamish, M., Venkataraghavan, R., McLafferty, F.W.: J. Am. Chem. Soc. 95, 3369 (1973).

470. Wittmann-Liebold, B., Lehmann, A.: In: Solid-Phase Methods in Protein Sequence Analysis. Laursen, R.A. (ed.). Rockford, Ill.: Pierce Chemical Co. 1975, p. 81.
471. Woods, K.R., Wang, K.T.: Biochim. Biophys. Acta 133, 369 (1967).
472. Wright, C.S., Alden, R.A., Kraut, J.: Nature (London) 221, 235 (1969).
473. Wuethrich, K.: In: NMR in Biological Research: Peptides and Proteins. New York: Elsevier 1976.
474. Wyckhoff, H.W., Tsernoglou, D., Hanson, A.W., Knox, J.R., Lee, B., Richards, F.M.: J. Biol. Chem. 245, 305 (1970).
475. Wyckhoff, H.W., Hardman, K.D., Allewell, N.M., Inagami, T., Johnson, L.N., Richards, F.M.: J. Biol. Chem. 242, 3984 (1967).
476. Yamada, S., Itano, H.A.: Biochim. Biophys. Acta 130, 538 (1966).
477. Young, M.A., Desiderio, D.M.: Anal. Biochem. 70, 110 (1976).

Subject Index

Vol. 17: Y.A. Vinnikov
Sensory Reception
Cytology, Molecular Mechanisms and Evolution
Translated from the Russian by W.L. Gray and B.M. Crook
124 figures (173 separate illustrations). IX, 392 pages. 1974
ISBN 3-540-06674-8

Vol. 18: H. Kersten, W. Kersten
Inhibitors of Nucleic Acid Synthesis
Biophysical und Biochemical Aspects
73 figures. IX, 184 pages. 1974
ISBN 3-540-06825-2

Vol. 19: M.B. Mathews
Connective Tissue
Macromolecular Structure and Evolution
31 figures. XII, 318 pages. 1975
ISBN 3-540-07068-0

Vol. 20: M.A. Lauffer
Entropy-Driven Processes in Biology
Polymerization of Tobacco Mosaic Virus Protein and Similar Reactions
90 figures. X, 264 pages. 1975
ISBN 3-540-06933-X

Vol. 21: R.C. Burns, R.W.F. Hardy
Nitrogen Fixation in Bacteria and Higher Plants
27 figures. X, 189 pages. 1975
ISBN 3-540-07192-X

Vol. 22: H.J. Fromm
Initial Rate Enzyme Kinetics
88 figures, 19 tables. X, 321 pages. 1975
ISBN 3-540-07375-2

Vol. 23: M. Luckner, L. Nover, H. Böhm
Secondary Metabolism and Cell Differentiation
52 figures, 6 tables. VI, 130 pages. 1977
ISBN 3-540-08081-3

Vol. 24: **Chemical Relaxation in Molecular Biology**
Editors: I. Pecht, R. Rigler
141 figures, 47 tables. XVI, 418 pages. 1977
ISBN 3-540-08173-9

Progress in Molecular and Subcellular Biology

Editors: F.E. Hahn, H. Kersten, W. Kersten, W. Szybalski. Advisors: T.T. Puck, G.F. Springer, K. Wallenfels. Managing Editor: R.E. Hahn

Volume 1
32 figures. VII, 237 pages. 1969
ISBN 3-540-04674-7

Volume 2
Proceedings of the Research Symposium on Complexes of Biologically Active Substances with Nucleic Acids and Their Models of Action
Held at the Water Reed Army Institute of Research, Washington, 16–19 March 1970
158 figures. IX, 400 pages. 1971
ISBN 3-540-05321-2

Volume 3
58 figures. VII, 251 pages. 1973
ISBN 3-540-06227-0

Volume 4
88 figures, 27 tables. XI, 251 pages. 1976
ISBN 3-540-07487-2

Volume 5
56 figures. XIII, 176 pages. 1977
ISBN 3-540-08192-5

Volume 5 of this series presents a collection of five progress reports concerning recent developments in molecular biology outside the classical area of molecular genetics. It continues the emphasis on medical pharmacological aspects of molecular biology which has characterized this series throughout its existence. The text covers the following topics in detail. F.E. Hahn: The Double Helix Revisted: Watson and Olby. C. Reiss, T. Arpa-Gabarro: Thermal Transition Spectroscopy: A New Tool for Submolecular Investigation of Biologic Macromolecules. I. Schuster: Interactions of Drugs with Liver Microsomes. W. Flamenbaum, J.H. Schwartz, R.J. Hamburger, J.S. Kaufman: The Pathogenesis of Experimental Acute Renal Failure: The Role of Membrane Dysfunction. P. Gund: Three-Dimensional Pharmacophoric Pattern Searching. N.W. Gabel: Chemical Evolution: A Terrestrial Reassessment.

Springer-Verlag Berlin Heidelberg New York

MIX
Papier aus verantwortungsvollen Quellen
Paper from responsible sources
FSC® C105338

If you have any concerns about our products, you can contact us on
ProductSafety@springernature.com

In case Publisher is established outside the EU, the EU authorized representative is:
Springer Nature Customer Service Center GmbH
Europaplatz 3, 69115 Heidelberg, Germany

Printed by Libri Plureos GmbH
in Hamburg, Germany